COLLINS GEM
BASIC FACTS

GEOGRAPHY

John Jilbert BSc

Revised by
Rosemary Hughes BA

COLLINS
London and Glasgow

First published 1983
Revised edition 1988
Reprint 10 9 8 7 6 5 4 3

ISBN 0 00 459106 2 (UK edition)
    0 00 459272 7 (Export edition)

Printed in Great Britain

Introduction
Basic Facts is a new generation of illustrated GEM dictionaries in important school subjects. They cover all the important ideas and topics in these subjects up to the level of first examinations.

Bold words in an entry mean a word or idea developed further in a separate entry: *italic* words are highlighted for importance.

Tables of important facts are placed together at the end of the dictionary.

Revised by Rosemary Hughes.

Abrasion The wearing away of the landscape by rivers, **glaciers**, the sea or wind, armed with a **load** of debris. See also **Corrasion**.

Abrasion platform See **Wave-cut platform**.

Accessibility A measure of the ease and efficiency with which a location can be reached. Central locations are highly accessible; peripheral ones not. Accessibility can be measured by the ß index.

Accessibility matrix A framework for calculating the **accessibility** of each **node** in a **network,** thus:

		TO						
		A	B	C	D	E	F	Row sum
	A	0	1	2	1	2	3	9
F	B	1	0	2	1	2	3	9
R	C	2	2	0	1	2	3	10
O	D	1	1	1	0	1	2	6
M	E	2	2	2	1	0	1	8
	F	3	3	3	2	1	0	12

Most accessible node = D (lowest row sum); least accessible node = F (highest row sum).

Acid rain Rain that contains a high concentration of pollutants, notably sulphur and nitrogen oxides.

1

These pollutants are produced from factories, power stations burning **fossil fuels** and car exhausts. Once in the **atmosphere** the sulphur and nitrogen oxides turn into sulphuric and nitric acids which fall as corrosive rain. The results of acid rain have been devastating to many lakes, forests and buildings in Scandinavia and Germany, due to **prevailing winds** in this part of the northern hemisphere blowing in a southwesterly direction.

Aeolian Relating to the wind: aeolian deposits are those transported and deposited by the wind. Hence aeolian **sandstone.** Wind **erosion** and **deposition** is at its most active in desert regions.

Afforestation The conversion of open land into a forest. This is usually for commercial reasons and often involves the planting of coniferous trees in upland areas.

Age and sex structure The classification of the elements of a national or regional population according to sex and age groups. Age and sex structure determines the shape of a **population pyramid.**

Agglomerate A mass of coarse róck fragments or blocks of lava produced during a volcanic erruption.

Agglomeration The tendency for firms with related products to locate in close proximity in order to reduce transport costs and other overheads. For example the printing and publishing industries, motor manufacture and component industries, oil refining and petrochemical industries.

Agriculture Human management of the **environment** to produce food. The numerous forms of agriculture can be categorized into three groups: **commercial agriculture,** where all produce is sold; **subsistence agriculture,** where all produce is consumed by the farming household; and **peasant agriculture,** where some produce is sold and some consumed.

Aid The provision of finance, personnel and equipment for furthering economic development and improving standards of living in the **Third World.** Most aid is organized by international institutions (e.g. the United Nations), by charities (e.g. Oxfam) or by national governments.

Alluvial fan A cone of **sediment** deposited at an abrupt change of slope; for example where a tributary post-glacial stream meets the flat floor of a **U-shaped valley.** Also common in arid regions where streams flowing off **escarpments** may periodically carry large **loads** of sediment during **flash floods.**

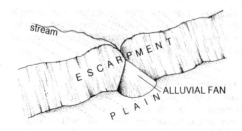

Alluvium Material deposited by a river in its middle and lower course. Alluvium comprises **silt,** sand and coarser debris eroded from the river's upper course and transported downstream. Alluvium is deposited in graded sequence: coarsest first (heaviest) and finest last (lightest). Regular floods in the lower course create extensive layers of alluvium which build up to considerable depth on the **flood plain.** Some of the world's most fertile lands are found on alluvial food plains.

Alp Gentle slope above the steep sides of a glaciated valley often used for summer grazing. See also **Transhumance.**

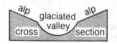

4

Amenity resources Those **resources** which provide an opportunity for recreation and leisure pursuits. The **national parks,** forests and the coast-line are good examples, as are areas of parkland and open space in cities.

Anemometer An instrument for measuring the velocity of the wind.

e.g. cup anemometer

Anthracite A hard form of **coal** with a high carbon content and few impurities.

Anticline An arch in folded **strata**; the opposite of **syncline.**

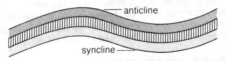

Anticyclone An area of high atmospheric pressure with light winds, clear skies and settled **weather.** In summer, anticyclones are associated with warm and sunny conditions; in winter they imply frost and fog as well as sunshine. On the weather map the anticyclone appears as on the diagram over page.

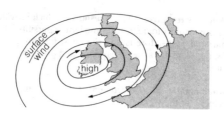

Appropriate technology Techniques and equipment which are appropriate to the immediate needs of a **developing country**. For example, ox-drawn planters and reapers, made of local materials, may be more useful in developing **agriculture** than tractors and combines, which are expensive to run, difficult to maintain in remote regions, and may cause unemployment. Appropriate technology stresses low cost, straightforward and **labour-intensive** projects as opposed to the **capital-intensive, high-technology approach**. See also **Intermediate technology**.

Aquifer See **Artesian basin**.

Arable farming The production of cereal and root crops — as opposed to the keeping of livestock.

Arête A knife-edged ridge separating two **corries** in a glaciated upland. The arête is formed by the progressive enlargement of corries by **weathering** and **erosion**. See also **Pyramidal peak**.

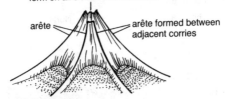

pyramidal peak developed where corries form on all sides of the mountain

arête

arête formed between adjacent corries

Arithmetic growth A rate of increase of the order of 2, 4, 6, 8, 10 . . . The concept is usually applied in the discussion of global food production compared with global **population growth.** Generally food production is increasing **exponentially.**

Artesian basin This consists of a shallow **syncline** with a layer of **permeable rock,** e.g. chalk, sandwiched between two layers of **impermeable rock,** e.g. clay. Where the permeable rock is exposed at the surface, rain water will enter the rock and the rock will become saturated. This is known as an *aquifer.* Boreholes can be sunk into the structure to trap the water in the aquifer. If there is sufficient pressure of water within the aquifer, the water will rise freely to the surface. The London Basin consists of a shallow syncline formed of a layer of chalk between two layers of clay. Because

London's **water table** has dropped considerably in recent years (due to demand by the Water Boards and industries) water now has to be pumped to the surface, as the pressure of water in the aquifer is too low for it to flow out freely.

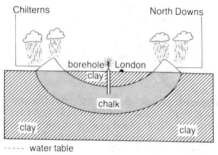

Chilterns

North Downs

borehole London

clay

chalk

clay

clay

----- water table

zone of saturation

Artificial fertilizer Chemical products containing one or all of nitrogen, potash, phosphates and trace elements. These are derived from petrochemicals, natural phosphate deposits and other industrial sources. Contrast with *natural fertilizer* — animal dung, rotted vegetation (compost) and animal derivatives such as bone meal.

Artificial fibres Textile materials made from hydrocarbons, e.g. nylon, rayon, viscose, Terylene made from such materials as oil, **coal,** wood fibres. Contrast with natural fibres wool and cotton.

Asymmetrical fold Folded **strata** where the two **limbs** are at different angles to the horizontal.

Assembly industry A firm which assembles components into a finished product is called an assembly industry. For example, the motor car industry assembles parts such as engines, bodies, wheels, windscreens and electrical components into a final product. The individual components are made by a number of other firms. Assembly industry is distinguished from, for example, a manufacturing industry such as steel-making which uses **primary** products — **coal, limestone,** iron ore.

Atmosphere The air which surrounds the earth: up to 15 km in thickness at the equator, less thick in higher latitudes. The atmosphere comprises Oxygen (21%), Nitrogen (78%), Argon, Helium and other gases in minute quantities.

Attrition The process by which the river's load is

eroded through particles such as pebbles and boulders striking each other.

Back-to-back A type of **terrace** house common in the industrial towns and cities of Britain in the late nineteenth and early twentieth centuries. Some still stand in the **inner cities**; many more have been demolished.

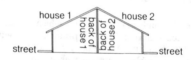

Backwash The return movement of seawater off the beach after a wave has broken. See also **Swash** and **Longshore drift.**

Barchan A type of sand dune, crescent-shaped, formed in desert regions where the wind direction is very constant. Wind blowing round the edges of the dune causes the crescent shape; the dune may advance in a down-wind direction due to particles being blown over the crest.

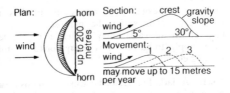

Bar graph A graph on which the values of a certain variable are shown by shaded columns; e.g.:

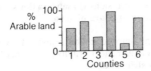

Barometer An instrument for measuring atmospheric pressure. There are two types, the mercury barometer and the aneroid barometer. The mercury barometer consists of a glass tube containing mercury which fluctuates in height as pressure varies. The aneroid barometer is a small metal box from which some of the air has been removed. This box expands and contracts as the air pressure changes. A series of levers joined to a pointer shows pressure on a dial.

11

Barrage A type of dam built across wide stretches of water, e.g. estuaries, for the purposes of water management. Such a dam may be intended to provide water supply, to harness wave energy or to control flooding etc.

Basalt A dark, fine-grained extrusive **igneous rock** formed when **magma** emerges onto the earth's surface and cools rapidly. The Giant's Causeway, Northern Ireland, is composed of basalt. As basalt cools a hexagonal pattern of jointing may occur as it contracts — hence the hexagonal basalt columns of Giant's Causeway and Fingal's Cave, island of Staffa, Scotland. A succession of basalt **lava flows** may lead to the formation of a **lava plateau.**

Basin The term commonly applied to the area drained by a river and its tributaries — hence 'drainage basin' or **catchment.**

Basin of internal drainage In desert regions depressions (sometimes below sea level) may occur, from which there is no natural outlet. Intermittent streams drain to the centre of the basin and **drainage**

occurs by evaporation. Lakes occupying such basins fluctuate in depth and area according to the balance of inflow and evaporation. Extensive salt deposits (e.g. halite, gypsum) may be left behind as lakes evaporate, e.g. Dead Sea, Makgadikgadi Pan (Botswana).

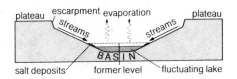

Batholith A large body of igneous material intruded into the earth's **crust**. As the batholith slowly cools, large-grained **rocks** such as **granite** are formed. Batholiths may eventually be exposed at the earth's surface by the removal of overlying rocks by **weathering** and **erosion**. Dartmoor and Exmoor in Southwest England are offshoots (bosses) from a batholith.

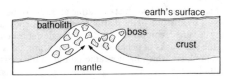

Bay An indentation in the coastline with a **headland** on either side. Its formation is due to the more rapid **erosion** of softer rocks.

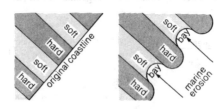

Bay bar A bank of sand or shingle extending as a barrier almost or totally across a **bay.** Caused by **longshore drift.**

Beach A strip of land sloping gently towards the sea, usually recognized as the area falling between high and low tide.

Sand is the final product of marine **erosion,** formed largely by **attrition** of the marine **load.** Beach deposits may be varied, comprising sand and shingle deposited by the sea, banks of pebbles deposited in storm tides, and boulders which have weathered out of the **cliffs** behind.

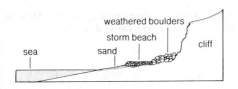

Bearing A compass reading between 0 and 360 degrees, indicating direction of one location from another, e.g.

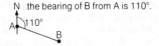

the bearing of B from A is 110°.

Bedding plane The division between two **stratas** of rock, generally indicating the boundary between earlier and later periods of **deposition.**

Bemba A Zambian people practising **bush fallowing.**

Bergschrund A large **crevasse** located at the rear of a **corrie** icefield in a glaciated region, formed by the weight of the ice in the corrie dragging away from the rear wall as the **glacier** moves downslope.

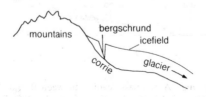

Bid-rent curve A type of graph used to indicate the rent-paying ability of land uses at different distances away from an urban centre. Bid-rent theory is central to both agricultural land use theories (**Von Thünen theory**) and models of urban structure (e.g. **Burgess model**). A family of bid-rent curves for an agricultural system would appear thus:

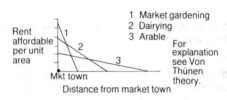

ß index (or **Beta index**) A measure of *connectivity* in a **network**. It is an expression of the ratio of **links** to **nodes**.

- node (settlement)   nodes = 5
— link (routeway)   links = 7
                          ß index = $\frac{7}{5}$ = 1·4

Biogas The production of methane and carbon dioxide, which can be obtained from plant or crop waste. Biogas is an example of a renewable source of energy (see **Renewable/nonrenewable resources**).

Birth rate The number of live births per 1000 people in a population per year.

Bituminous coal Sometimes called housecoal — medium quality **coal** with some impurities; the typical domestic coal. Also the major fuel source for **thermal power stations**.

Blackband iron ore Ore found interbedded with **coal** deposits in many of Britain's coalfields. This was the earliest source of raw material for the growing iron smelting industry of the **industrial revolution**.

Blast furnace Chamber in which an air blast causes coke to burn at a sufficiently high temperature to allow the smelting of iron from iron ore in the presence of **limestone**, a **flux**.

17

Block faulting The dissection of a region by an extensive system of vertical or semi-vertical **faults**. The faults divide the landscape into a series of blocks which may be relatively uplifted or depressed, to produce a series of **block mountains** (horsts) and **rift valleys**.

Block mountain (or **horst**) A section of the earth's **crust** uplifted by faulting. Mt. Ruwenzori in the East African Rift System is an example of a block mountain.

Block train A goods train comprising a single freight type, such as ore, grain or oil. Contrast with the general freight train which carries a variety of goods in different types of wagons.

Blowhole A crevice, **joint** or **fault** in coastal rocks produced by marine **erosion**, often leading from the rear of a cave up to the **cliff** top. As waves break in the cave, air and sometimes spray are forced up the blowhole to erupt at the surface.

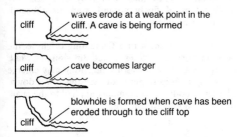

waves erode at a weak point in the cliff. A cave is being formed

cave becomes larger

blowhole is formed when cave has been eroded through to the cliff top

Blue-collar worker A worker who is either a manual worker or who works in potentially 'dirty conditions'. The term 'blue-collar' derives from the idea of wearing dark-coloured overalls for work (compare **white-collar worker**).

Blue ice Heavily compressed ice at the bottom of a **corrie** icefield or **glacier**: air has been expelled by compression and the ice therefore appears blue. Contrast with **white ice**.

Bluff See **river cliff**.

Boulder clay (or **till**)   The unsorted mass of debris dragged along by a **glacier** as *ground moraine* and dumped as the glacier melts. Boulder clay may be several metres thick and may comprise any combination of finely ground 'rock flour', sand, pebbles or boulders.

Bourne   A stream with a fluctuating origin as the position of **springs** varies on the **dip slope** of **chalk escarpments**. In wet seasons the spring may emerge higher up the dip slope; in dry seasons lower down. The term 'bourne' is commonly applied to the chalk regions of Southern England.

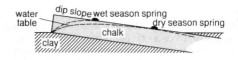

BP   abbrev. for Before Present, a term used in geological time scales to denote any time before the present.

Break-of-bulk-point   A situation in freight carriage where goods are off-loaded from one transport medium and loaded onto another. A port is a major break-of-bulk-point, as is a railhead. Minimizing break-of-bulk-points is one way of reducing costs in transport systems.

Breakwater (or **groyne**) A wall built at right angles to a beach in order to prevent sand loss due to **longshore drift**.

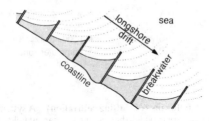

Breccia Rock fragments cemented together by a matrix of finer material: the fragments are angular and unsorted, e.g. volcanic breccia which is made up of coarse angular fragments of **lava** and **crust** rocks welded by finer material such as ash and **tuff**.

21

Burgess model (or **concentric theory**) A model of urban structure formulated by E. W. Burgess in 1923 and based on Chicago, in which five major zones of uban land use are proposed:

1: central business district
2: transition zone (factory zone)
3: workingmen's homes
4: residential zone
5: commuter zone

Bush fallowing (or **shifting cultivation**) A system of **agriculture** in which there are no permanent fields: for example in the **tropical rain forest**, remote societies cultivate forest clearings for one year and then move on.

The system functions successfully when forest **regeneration** occurs over a sufficiently long period to allow the soil to regain its fertility. A typical bush fallowing régime might appear as on the opposite page.

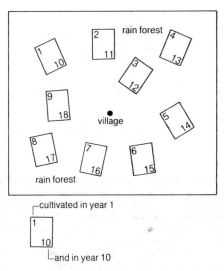

Butte An outlier of a **mesa** in arid regions.

Caldera A large crater formed by the collapse of the summit cone of a **volcano** during an eruption. The caldera may contain **subsidiary cones** built up by subsequent eruptions, or a crater lake if the volcano is extinct or dormant. See diagram over page.

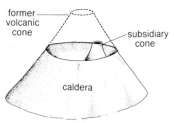

former volcanic cone

subsidiary cone

caldera

e.g. Crater Lake, Oregon, USA

Canyon A deep and steep-sided river valley occurring where rapid vertical **corrasion** takes place in arid regions. In such an **environment** the rate of **weathering** of the valley sides is slow. If the **rocks** of the region are relatively soft then the canyon profile becomes even more pronounced. The Grand Canyon of the Colorado River in the USA is the classic example. See diagram on opposite page.

Capital intensive An operation in which high productivity is achieved through high investment; e.g. **market gardening** is a capital intensive form of **agriculture** identified by high investment in equipment such as greenhouses and heating and irrigation systems. See also **Labour intensive**.

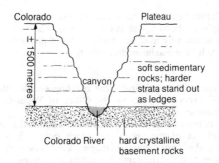

Colorado

Plateau

± 1500 metres

canyon

soft sedimentary
rocks; harder
strata stand out
as ledges

Colorado River   hard crystalline
basement rocks

Cap rock (or fall maker) A stratum of resistant **rock** at the lip of a **waterfall**.

Catchment Physically, an alternative term to **basin**; in **human geography** catchment refers to an area around a town or city — hence 'labour catchment' to describe the area from which an urban workforce is drawn.

Cavern In **limestone** country, a large underground cave formed by dissolving of limestone by subterranean streams. See also **Stalactite, Stalagmite**.

CBD The Central Business District. This is the central zone of a town or city, and is characterized by high **accessibility**, high land values and limited space.

The visible result of these factors is a concentration of high-rise buildings at the city centre. The CBD dominated by retail and business **functions**, both of which have maximum accessibility.

Census A questionnaire sent to all **households** in the UK every ten years by the Office of Population Censuses and Surveys.

Information requred by the census includes: the age and sex of the occupants; which people in the household work; where do they work and how do they travel to work; the number of rooms in the house; the nature of the **household amenities**, and whether the house has a garage. This information is used to assess future needs of particular areas and households, e.g. whether new schools or old people's homes are likely to be required in the future. It is illegal not to fill in the census.

Central place A **settlement**, offering goods and **services** to a surrounding population.

Central place theory Walter Christaller's 1933 model of **settlement** location and distribution. Christaller envisaged a **settlement hierarchy** in which small **central places** would offer a limited range of everyday goods and **services** to a small surrounding population, and large central places would offer a large number of goods and services, many of them of a specialized nature, to a large surrounding population. Thus the **sphere of influence** of the large central places would be considerably greater in extent than that of the small central places. Christaller proposed several orders of central places, and in an idealized and uniform **environment** a typical Christaller hierarchy would appear thus:

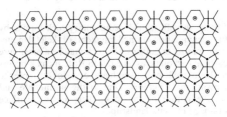

- • first order central place

- ⬡ first order sphere of influence

- ⊙ second order central place

- ⬡ second order sphere of influence

  further orders can be added: 3, 4 . . .

Centrality index A measure of the status of a **central place** in the **settlement hierarchy**, as calculated by

$$C = \frac{t}{T} \times 100.$$

Where:
C is the centrality index for an outlet of a given function.
t is 1.
T is the total number of outlets of the function in a given settlement system.

C is calculated for all **functions** in a given **settlement** in order to determine its total centrality index.

Chalk A soft, whitish **sedimentary rock** formed by the accumulation of small fragments of skeletal matter from marine organisms: the rock may be almost pure calcium carbonate as a result. In Britain chalk occurs in the low hills of the south and east, for example the Downs, the Chilterns. Due to the **permeable** and soluble nature of the rock, there is little surface **drainage** in chalk landscapes.

Chernozem A deep, rich soil of the plains of the southern USSR. The upper **horizons** are rich in lime and other plant nutrients; in the dry **climate** the predominant movement of **soil** moisture is upwards (contrast with *leaching*), and bases therefore accumulate in the upper part of the **soil profile**.

Chitamene A system of **bush fallowing** practised by the Bemba people of Zambia.

Cirrus High, wispy or strand-like thin cloud associated with the advance of a **depression**.

Clay A **soil** composed of very small particles of **sediment**, less than 0.002 mm in diameter. Due to the dense packing of these minute particles, clay is almost totally impermeable, i.e. does not allow water to drain through. Clay soils very rapidly waterlog in wet **weather**.

Cliff A steep rockface between land and sea; the profile of the cliff will be determined largely by the nature of the coastal rocks. Resistant rocks such as **granite** (e.g. Land's End) will produce steep and rugged cliffs. The dip of the **strata** will influence the gradient of the cliff. See also **Blowhole**.

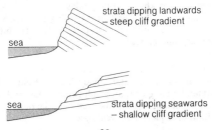

Climate The average atmospheric conditions prevailing in a region; distinct from **weather** in that a statement of climate is concerned with long-term trends. Thus the climate of, for example, the Amazon Basin is described as hot and wet all the year round; that of the Mediterranean Region as hot dry summers and mild wet winters. See **Extreme climate, Maritime climate**.

Clint A block of **limestone**, especially as part of a **limestone pavement**, where the surface is composed of clints and **grykes**.

Coal A **sedimentary rock** composed of decayed and compressed vegetative matter; coal in Britain dates from the Carboniferous period (about 350 million years ago) during which time tropical forest covered large areas of what is Britain today.

Coal is usually classified according to a scale of hardness and purity: **anthracite** (hardest), **bituminous, lignite, peat**.

Coastal marsh as at Romney Marsh, Kent, UK; formed by the growth of a **spit** across a **bay**, and the gradual silting up of the resulting **lagoon**. The following sequence of events occurs:

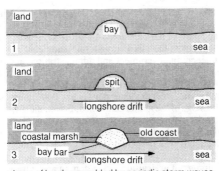

shape of bay bar moulded by periodic storm waves

The coastal marsh is consolidated in the following sequence:
a) **Siltation** – formation of mudflats covered at high tide
b) **Salt marsh** – colonization by reeds and other plants
c) **Freshwater marsh** – plants and soil build up above high water.

Coking coal High-quality **coal** used in the **blast furnaces** of the iron and steel industry.

Cold front See **Depression.**

31

Colonial influence A term describing the consequences of, for example, British colonial activity in many parts of Africa. There are two broad areas where colonialism has had long-term and continuing influence: in government and administration, and in economic systems. Many former colonies have governments and bureaucracies modelled on British institutions, and this may not always be to the best advantage of the nation concerned. Economically, the majority of former colonies continue to be suppliers of **primary products** to the industrial world, a situation which may well perpetuate under-development.

Commercial agriculture A system of **agriculture** in which food and materials are produced specifically for sale in the market — contrast with **subsistence agriculture**. Commercial agriculture tends to be **capital intensive**.

Common land Land which is not in the ownership of an individual or institution, but which is historically available to any member of the community. Hence common grazing rights still obtain in some upland areas. The village green, where it survives, is often common land.

Communications The contacts and linkages in an **environment**. For example roads and railways are communications, as are telephone systems, newspapers and broadcasts.

Commuter zone   An area on or proximal to the outskirts of an urban area. Commuters are the most affluent and mobile members of the urban community and can afford the greatest physical separation of home and work. The commuter zone is thus the outer ring of suburbs (see **Burgess model**) and the villages beyond, the latter increasingly peopled by managers, professionals, etc. who travel daily to the city centre. See **Dormitory settlement**.

Concentric theory   See **Burgess model**.

Concordant coastline   A coastline that is parallel to mountain ranges immediately inland. A rise in sea level or a sinking of the land will cause the valleys to be occupied by the sea and the mountains to become a line of islands. This has occurred along the coast of Yugoslavia in the Adriatic sea. Compare **Discordant coastline**.

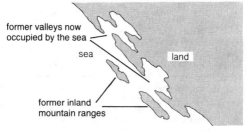

former valleys now
occupied by the sea

sea

land

former inland
mountain ranges

Conflicting and compatible demands   In resource use, the key management problem is the resolution of different demands for a single resource. For example in water use, some demands are conflicting (sewage dilution and domestic supply); some compatible (recreation and flood control). Ideally users should be distributed such that no single use impairs the quality or quantity of the resource for any subsequent user.

Concealed coalfield   Where **coal** measures are located at some depth beneath overlying **strata**, a concealed coalfield exists. Note also *exposed coalfield* where coal measures outcrop at or near the surface, e.g. Yorkshire coalfield:

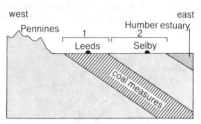

1 Exposed coalfield – open cast or shaft mining since late eighteenth century.
2 Concealed coalfield – recent development of deep mining. (Inclined shafts or drifts.)

Coniferous forest For example, the **taiga** of northern Asia, lying between the **tundra** to the north and the steppes to the south. The taiga is composed of trees such as pine, spruce, fir interspersed with large areas of swamp and **flood plain**. Coniferous woodland occurs considerably further north than broadleaved deciduous species; the biology of the conifer will withstand much harsher climatic conditions. The majority of conifers are **evergreen**; some, like the larch, are **deciduous.**

Connectivity See ß index.

Conservation The preservation and management of the natural **environment**. In its strictest form, conservation may mean total protection of endangered species and habitats, as in nature reserves. In some cases, conservation can refer to the artificial environment, e.g., the conservation of ancient buildings.

Continental Drift The theory that the earth's continents move gradually over a layer of semimolten rock underneath the earth's **crust**. It is thought that the present day pattern of continents is derived from the supercontinent **Pangea** which existed approximately 200 million years ago. See also **Gondwanaland, Laurasia**.

Contour A line drawn on a map to join all places at the same height above sea level. Contour heights are expressed in metres on British Ordnance Survey maps (e.g. 1:50,000 series), and the interval between contours is usually 10 metres (1:50,000 series) or 5 metres (1:25,000 series). Contours are generally shown in brown.

Contour ploughing A method of soil **conservation** whereby ploughing is undertaken along **contours** rather than with the slope. The effect of this strategy is to reduce the rate of runoff and thus to retain **soil** that would otherwise be washed off downslope. The risk of sheet erosion and gully erosion (see **soil erosion**) is thus reduced.

Contract farming A system in which individual farmers contract with food processing firms to produce a specific crop. The firm will often provide seed and other inputs at the start of the season, and will purchase the entire crop at harvest time. Contract farming is common in East Anglia, where firms such as 'Birds Eye' are operating.

Conurbation A continuous built-up urban area, formed by the merging of several formerly separate towns or cities. Hence, for example, the West Yorkshire conurbation comprising Leeds, Bradford, Wakefield, Huddersfield, Halifax and others. Several stages can be identified in the growth of conurbations:

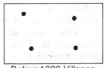
Before 1800. Villages

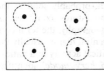
1800 to 1900.
Rapid industrialization
and urbanization

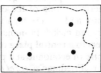
After 1900.
Towns merge into a
continuous urban
area – a conurbation

----- limit of built-up area

Evidence of these processes can be observed, for example, on the Yorkshire coalfield where small rural **settlements** grew dramatically during the nineteenth century as steam power gave rise to the expansion of mining and manufacturing industry. Twentieth-century **urban sprawl** has led to the merging of towns.

Cooperative   A cooperative farming system is one in which individual farmers pool their **resources** in order to optimize individual gains.
   Cooperative purchasing, e.g. of machinery, will provide a farmer with shared use of a capital investment which an individual could not normally afford. Cooperative marketing is similarly more efficient than individual selling.

Core   In **physical geography**, the core is the innermost zone of the earth, probably solid at the centre and at very high temperatures, composed of iron and nickel. In **human geography**, the term core refers to a **central place** or central region, usually the centre of economic and political activity in a nation. John Friedman's *core/periphery model* identifies relationships between growing and stagnant regions during the process of economic development as follows:

Pre-industrial phase: independent villages with local **spheres of influence**.

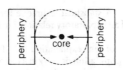

Industrializing phase: a core region emerges, based on a growing industrial urban centre — the rest of the nation remains an undeveloped **periphery**, supplying labour, food and other **resources** to the core.

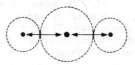

Fully developed phase; regional urban centres develop to spread the benefits of economic and social progress to the former peripheries.

This is a highly simplified version of a complex four-stage model of regional development.

Corrasion   The abrasive erosive action of an agent of **erosion** (rivers, ice, the sea) through its **load**; for example the pebbles and boulders carried along by a river wear away the channel bed and sides.

Correlation A statistical technique for establishing the extent of an exploratory relationship between two variables.

In search of an explanation of, for example, variations in **population density** in a region, one might consider factors such as altitude, **accessibility** and resource indices as possible explanatory variables. In such a case population density is regarded as the *dependent variable*; and the other factors as *independent variables*. The dependent variable (y) is correlated with the independent variable (x) in order to measure the strength of the relationship, ranging on a scale −1 to +1. Zero implies no explanatory relationship; −1 a perfect inverse relationship; +1 a perfect direct relationship. There are two commonly used correlation statistics in geography: the *Spearman Rank Correlation Coefficient*:

$$R = 1 - \left[\frac{6\sum d^2}{n^3 - n} \right]$$

and the *Pearson product Moment Correlation Coefficient*:

$$r = \frac{\sum(x - \bar{x})(y - \bar{y})/n}{\sigma x \, \sigma y} \quad \text{(where } \sigma \text{ is standard deviation).}$$

See tables on opposite page and example on page 42.

Tables needed for Spearman:

District	x	y	Rank x	Rank y	Difference in rank (d)	d^2
1						
2						
3						
n						
					$\sum d^2 =$	

Tables needed for Pearson:

District	x	$x - \bar{x}$	$(x - \bar{x})^2$	y	$y - \bar{y}$	$(y - \bar{y})^2$	$(x - \bar{x})(y - \bar{y})$
1							
2							
n							
	$\bar{x} =$		$\sum =$	$\bar{y} =$		$\sum =$	$\sum =$

41

An example of Spearman Rank Correlation:
Data: (mid 1970's)

Region	Per capita income ($)	Rate of population growth (% per year)
1 West Africa	300	2.6
2 S. E. Asia	260	2.4
3 Tropical South America	960	2.8
4 North America	7020	0.5
5 Western Europe	6150	0.1
6 Southern Europe	2470	0.8

Region	x	y	Rank x	Rank y	Diff. in rank (d)	d^2
West Africa	300	2.6	5	2	3	9
S. E. Asia	260	2.4	6	3	3	9
Tropical South America	960	2.8	4	1	3	9
North America	7020	0.5	1	5	4	16
Western Europe	6150	0.1	2	6	4	16
Southern Europe	2470	0.8	3	4	1	1

$$\sum d^2 = 60$$

$$R = 1 - \left[\frac{6\Sigma d^2}{n^3 - n} \right]$$

$$= 1 - \left[\frac{6 \times 60}{6^3 - 6} \right]$$

$$= \underline{-0.7}$$

Interpretation:
There is a strong inverse relationship between standard of living (**per capita income**) and rate of **population growth**; i.e. as incomes rise, population growth slows down. See **Demographic transition theory** for detailed explanation.

Corrie (or **cirque** or **cwm**) A bowl-shaped hollow on a mountainside in a glaciated region; the area where a valley **glacier** originates. In glacial times the corrie contains an icefield, which in cross section appears as follows:

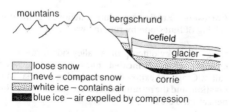

The shape of the corrie is determined by the rotational erosive force of ice as the glacier moves downslope:

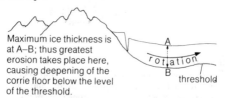

Maximum ice thickness is at A–B; thus greatest erosion takes place here, causing deepening of the corrie floor below the level of the threshold.

After the glacial period the corrie may contain a **tarn** fed by post-glacial streams.

Corrosion   **Erosion** by solution action; e.g. the dissolving of **limestone** by running water.

Crag   A rocky outcrop on a valley side formed, for example, when a **truncated spur** exists in a glaciated valley. Crags are exposed to **weathering** especially by **nivation**, and the resulting debris accumulates as **scree** beneath the crag.

Crag and tail A feature of lowland **glaciation**, where a resistant rock outcrop withstands **erosion** by a **glacier** and remains as a feature after the **ice age**. Rocks of volcanic or metamorphic origin are likely to produce such a feature. As the ice advances over the crag, material will be eroded from the face and sides and will be deposited as a mass of boulder clay and debris on the leeward side, thus producing a 'tail'.

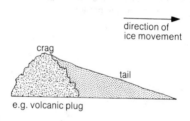

e.g. volcanic plug

Example of crag and tail: Castle Rock, Edinburgh.

Crater lake A lake occupying a **caldera** of a dormant or extinct **volcano**.

Credit Financial or material loan to allow investment, for example in agricultural development in the **Third World**. The loan would be repaid once profits have been made.

Crevasse A crack or fissure in a **glacier** resulting from the stressing and fracturing of ice at a change in **gradient** or valley shape. Crevasses range from small surface cracks to major factures many metres in depth, and may occur at any angle although two main orientations are recognized: transverse and longitudinal. Transverse crevasses occur where there is a change of gradient in the valley floor; longitudinal where the valley becomes wider and the ice mass stretches to occupy the broader space.

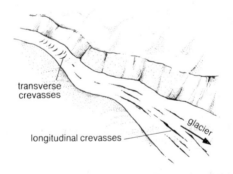

transverse crevasses

longitudinal crevasses

glacier

Cross section In mapwork, a cross section depicts the **topography** of a system of **contours**, as in the diagram on the opposite page.

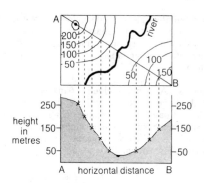

Crust (or **lithosphere**) The outermost layer of the earth. The solid crust rides on the fluid **mantle** and comprises two layers — the ocean crust and the continental crust. The crust is fractured into a series of plates. See also **Plate tectonics**.

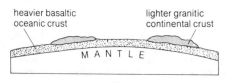

heavier basaltic oceanic crust

lighter granitic continental crust

MANTLE

Cumulonimbus A heavy dark cloud of great vertical depth. It is the typical thunderstorm cloud, giving heavy showers of rain, snow or hail. Such clouds form where intense solar radiation causes vigorous convection.

Cumulus A large cloud (smaller than a **cumulonimbus**) with a 'cauliflower' head and almost horizontal base. It is indicative of fair, or at worst, showery **weather** in generally sunny conditions. The base of the cumulus layer indicates *condensation level* in a rising and cooling body of moist air.

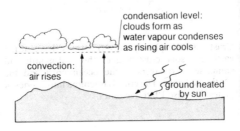

condensation level:
clouds form as
water vapour condenses
as rising air cools

convection:
air rises

ground heated
by sun

Cut-off See **Oxbow lake.**

Cycle of erosion The classic theories of landscape evolution envisaged a sequence of events whereby a newly uplifted land mass would be worn down by **weathering** and the agents of **erosion** (rivers, ice, sea, wind) to produce a flat plain (peneplain) with residual low hills. This latter would then be uplifted by earth movements and the cycle would begin again.

Dairying **Pastoral farming** system producing milk and related products such as cheese, butter, cream and yoghurt.

Death rate The number of deaths per 1000 people in a population per year.

Deciduous woodland Trees which are generally of broadleaved rather than coniferous habit, and which shed their leaves during the cold season. The larch is a deciduous conifer and is thus the exception to the **evergreen** norm in such trees.

Deflation In desert regions the removal of loose sand by wind **erosion**, often to expose a bare rock surface beneath, is referred to as 'deflation'.

Deforestation The practice of clearing trees. Much deforestation is a result of development pressures, e.g. trees are cut down to provide land for agriculture and industry. Deforestation in some **Third World** countries has led to severe **soil erosion**, a consequence of which has been **desertification** and eventually famine. It is thought that many of the famines in African countries during recent years have been partly caused by deforestation.

Delta A fan-shaped mass consisting of the deposited **load** of a river where it enters the sea. A delta only forms where the river deposits material at a faster rate than can be removed by coastal currents. While deltas may take almost any shape and size, three types are generally recognized:

Arcuate delta
e.g. Nile
Note bifurcation of river into distributaries in delta.

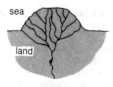

Bird's foot delta
e.g. Mississippi

Estuarine delta
e.g. Amazon

50

Demographic transition theory A model of **population change** which suggests the following pattern of changes over time:

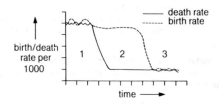

1) **Pre-industrial** societies: high **death rate** due to factors such as low levels of technology and vulnerability to natural disaster. High **birth rate** to ensure survival of population.

2) Industrializing societies: improvements in, e.g., nutrition, **communications** and medical facilities, have an early effect on death rate which falls rapidly. Birth rate continues high.

3) Urban/industrial societies: in highly developed economies the majority of the population is urban and in paid employment; as a result birth rate falls. Small families are likely owing to high mobility in the population, and the passing of the traditional necessity of a large family (= labour) in an agricultural society.

The theory can be seen as the history of one nation's population evolution, or a classification for a number of different nations at various stages in the development process. Although the model is a very generalized and simplified form of reality, it does emphasize the essential link between **population change**, economic development and **urbanization**.

Denudation The general term for the wearing away of the earth's surface by the processes of **weathering** and **erosion**.

Deposition The general term for the laying down of **sediments** resulting from **denudation**.

Depressed region An area of economic stagnation or decline, characterized by high unemployment, out-migration, declining private and public investment. Depressed regions are often sites of traditional **heavy industry** such as steel-making and engineering; as the **resource** base for such activity contracts, and as demand for their products falls, so the regional economy decays. Parts of Northeast England, Merseyside and Clydeside are depressed regions in Britain. Central government attempts to revitalize depressed areas through **development area** policy.

Depression An area of low atmospheric pressure

occurring where warm and cold air masses come into contact, for example in the case of the northern hemisphere along the north polar front (30°−60°N) where prevailing southwesterly winds (bringing moist, warm tropical air northwards) meet prevailing northeasterly winds (bringing cold polar air southwards). The passage of a depression is marked by thickening cloud, rain, a period of dull and drizzly weather and then clearing skies with showers. A depression develops as follows (see also diagram over page):

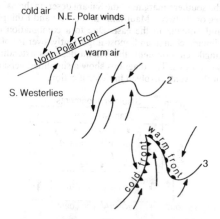

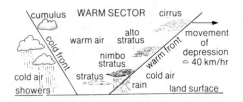

Desertification The encroachment of desert conditions into areas which were previously once productive. The problem is especially serious along the southern margins of the Sahara desert in the Sahel region, between Mauritania in the west and Ethiopia and Somalia in the east. Due to a combination of climatic change and, most importantly, overuse of a fragile **environment**, the Sahara desert is extending southwards. The diagrams show a simplified version of the events involved in desertification.

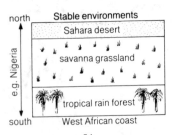

54

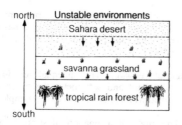

north Unstable environments

Sahara desert

savanna grassland

tropical rain forest

south

The extension of desert conditions into the **savanna** zone is made worse by overgrazing of limited pasture and by the lowering of the **water table**, due to the sinking of large numbers of boreholes for watering stock.

Developing countries   A collective term for those nations in Africa, Asia and Latin America which are undergoing the complex processes of modernization, **industrialization** and **urbanization**. There is great political and social variation between countries loosely grouped as 'developing' and such terms should be used with caution.

Development area   In Britain, a region designated by government as in need of special assistance for economic reconstruction. See **Depressed region**.

Development process   The sequence of events by

which a nation moves from a predominantly subsistence agricultural economy to one based on **commercial agriculture**, industry and a highly urbanized society.

Differential erosion The unequal **erosion** of interbedded hard and soft rocks, the softer **strata** being worn away more quickly. See, for example, **Bay**.

Dip The angle of inclination of **strata** from the horizontal, e.g.:

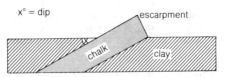

$x° = dip$

escarpment

chalk

clay

Dip slope The gentler of the two slopes either side of an escarpment crest; the dip slope trends in the direction of the dipping **strata**; the steep slope in front of the crest is the **scarp slope**.

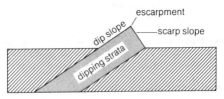

escarpment

dip slope

scarp slope

dipping strata

Discharge   The volume of runoff in the channels of a river **basin**. Discharge is expressed as follows:

$$Q = AV$$

Where:
Q is discharge in cubic metres per second.
A is the cross-sectional area of the channel.
V is the mean velocity of the stream.

Discordant coastline   A coastline that is at right angles to the mountains and valleys immediately inland. A rise in sea level or a sinking of the land will cause the valleys to be flooded. A flooded river valley is known as a **rig** whilst a flooded glaciated valley is known as a **fiord**. Compare **Concordant coastline**.

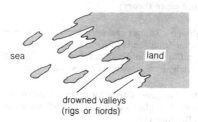

sea

land

drowned valleys
(rigs or fiords)

Diversification   A broadening of an agricultural or industrial product range in order to reduce dependence on a single, perhaps vulnerable, product. Diversification in **agriculture** is ecologically sound since a more varied **ecosystem** will have a healthier pest-predator complex: uniform wheatfields, for example, are susceptible to explosions of plant-specific pests which will require expensive (and perhaps pollutive) artificial control. Such **mono-culture** also progressively drains the **soil** of specific nutrients. Diversification is economically sound as an insurance against falling markets for a single product.

Dormitory settlement A village located beyond the edge of a city but inhabited by residents who work in that city (see **commuter zone**). The population of dormitory **settlements** is disproportionately large for the number of goods and **services** available in it (see **central place theory**).

Drainage The general term applied to the removal of water from the land surface by processes such as streamflow and infiltration. Drainage can be hastened artificially by the laying of pipes and culverts.

Drift The general term for material transported and deposited by glacial action of the earth's surface. See also **Boulder clay**.

Drift mine A system of mining in which an inclined plane gives access to the ore. In the newer areas of the Yorkshire coalfield based on Selby, inclined roadways lead to the coalface, allowing free access for plant and machinery.

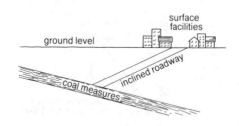

Dry valley (or **coombe**) A feature of **limestone** and **chalk** country, where valleys have been eroded in what are today dry landscapes. Such valleys may date from a period of more moist **climate** or from the end of the last **glaciation** when *periglacial* conditions, specifically a frozen subsoil and bedrock, sealed the otherwise **permeable** limestone and thus surface streams existed.

Dyke 1. An artificial **drainage** channel.

2. An artificial bank built to protect low-lying land from flooding.

3. A vertical or semi-vertical igneous intrusion occurring where a stream of **magma** (e.g. from a

batholith) moves through a line of weakness in the surrounding **rock**.

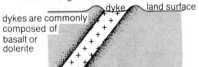

Cross section of eroded dyke, showing how metamorphic margins, harder than dyke or surrounding rocks, resist erosion.

dyke — land surface

dykes are commonly composed of basalt or dolerite

Metamorphosed zone: surrounding rocks close to intrusion are 'baked'.

Eastings The first element of a **grid reference**, as in the following example:

The bottom left corner of the map is taken as the origin; eastings are read towards the right edge of the map, northings towards the top edge.

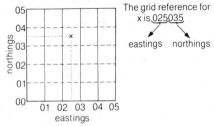

The grid reference for x is 025035

eastings   northings

Ecology The study of living things, their inter-relationships and their relationships with the **environment**.

Economies of scale The savings made in industry through mass production, automation and integrated processes. The unit cost of products falls as the quantity manufactured increases; investment in sophisticated manufacturing processes can further reduce unit costs.

Ecosystem A natural system comprising living organisms and their **environment**. The concept can be applied at the global scale or in the context of a smaller defined environment, e.g. a woodland, a pond or a marsh. Whatever the scale, the principle of the ecosystem is constant: all elements are intricately linked by flows of energy and nutrients, and a change in one element will have effects on the rest of the system.

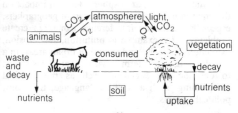

Employment structure The distribution of the workforce between the **primary, secondary, tertiary** and **quaternary sectors** of this economy. Primary employment is in **agriculture**, mining, forestry and fishing; secondary in manufacturing; tertiary in the retail, service and administration category; quaternary in information and expertise. One way of looking at the level of development of a nation is to examine its employment structure. Most employment in the **Third World** is primary, while that in the more developed nations is tertiary, with an increasing number of people being employed in the quaternary sector.

Enterprise zone An area, usually in the inner city, where special grants and facilities are made available for reconstruction and development.

Environment Human physical surroundings: **soil**, vegetation, wildlife, **atmosphere**. Human impact on the environment is a major concern of geographers, especially as human interference can often create problems — for example in **pollution, soil erosion**, extinction of species and spread of urban areas. In a broader sense, the term environment is also used to describe the social as well as the physical surroundings such as culture, language, traditions, political systems.

Erosion The wearing away of the earth's surface by running water (rivers and streams), moving ice (**glaciers**), the sea and the wind. These are called the *agents* of erosion.

Erratic A boulder of a certain rock type resting on a surface of different geology. For example at Norber Blocks in Yorkshire, blocks of **granite** rest on a surface of carboniferous **limestone**.

The explanation is glacial: the erratics were picked up by **glaciers** moving southwards over Britain and deposited at the point where the ice melted. Erratic blocks of this kind may be found at considerable distances from their origin.

Escarpment A ridge of high ground as, for example, in the **chalk** escarpments of southern England (The Downs, The Chilterns). See diagram over page.

Esker A low, winding ridge of pebbles and finer **sediment** on a glaciated lowland. An esker marks the course of a sub-glacial stream, flowing as melt water increases in volume towards the margin of a lowland ice body. As the stream flows it deposits sediment along its course, and this is left behind as a landscape feature after the end of the glacial period.

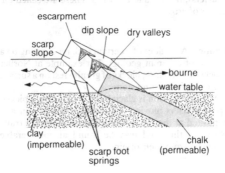

simplified block
diagram of a
chalk escarpment

escarpment

dip slope   dry valleys

scarp
slope

bourne

water table

clay
(impermeable)

scarp foot
springs

chalk
(permeable)

Estuary The broad mouth of a river where it enters the sea. An estuary forms where opposite conditions to those favourable for **delta** formation pertain: deep water offshore, strong marine currents, smaller **sediment** load (perhaps due to **deposition** upstream, e.g. in a lake).

Evaporite A type of **sedimentary** rock formed where salts are precipitated by evaporation in hot climates. Typical evaporites are gypsum and halite (rock salt), which are formed around the margins of lakes in **basins of internal drainage**, e.g. the Dead Sea and Sea of Galilee.

Evergreen Vegetation type in which leaves are continuously present. (Compare **deciduous woodland**).

Evapotranspiration The return of water vapour to the **atmosphere** by evaporation from land and water surfaces and the **transpiration** of vegetation.

Exfoliation A form of **weathering** whereby the outer layers of a **rock** or boulder sheer off due to the alternate expansion and contraction produced by diurnal heating and cooling. Such a process is especially active in desert regions where, due to clear skies, night temperatures drop considerably below daytime peaks. Isolated boulders may be surrounded by exfoliation debris. See diagram over page.

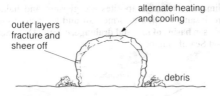

alternate heating and cooling

outer layers fracture and sheer off

debris

Exports The selling of goods and services to a foreign country (compare **imports**).

Exposed coalfield See **Concealed coalfield**.

Exponential growth A sequence, for example 2, 4, 8, 16, 32 . . . is termed exponential growth, as contrasted to **arithmetic growth** 2, 4, 6, 8, 10, etc. World population is often quoted as an example of exponential growth:

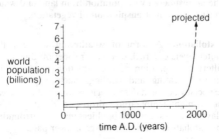

projected

world population (billions)

time A.D. (years)

Extensive farming   A system of **agriculture** in which relatively small amounts of capital or labour investment are applied to relatively large areas of land. For example sheep ranching is an extensive form of farming, and yields per unit area are low. Extensive farming usually occurs at the *margin* of the agricultural system — at great distance from market or on poor land of limited potential.

External processes   Landscape-forming processes such as **weather** and erosion; contrast with **internal processes**.

Extreme climate   For example in continental interiors: in such locations there is a large annual range in temperature (and in some cases rainfall). In Central Asia hot summers alternate with very cold winters, and rainfall is concentrated as summer (convectional) maximum. Such a **climate** is regarded as extreme. Compare **Maritime climate**.

Fault   A fracture in the earth's **crust** either side of which the **rocks** have been relatively displaced. The scale of faulting can vary considerably, from a small surface crack to a fracture of regional extent. Faulting occurs in response to stress in the earth's crust: the release of this stress in fault movement is experienced as earthquakes. There are three broad categories of fault:

tear fault

lateral movement

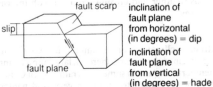

fault scarp

slip

fault plane

inclination of fault plane from horizontal (in degrees) = dip

inclination of fault plane from vertical (in degrees) = hade

Fault terminology can be summarized as follows:

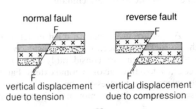

normal fault

F

F

vertical displacement due to tension

reverse fault

F

F

vertical displacement due to compression

Fell Upland rough grazing in a **hill farming** system, for example in the English Lake District. Fell land is sometimes **common land**, i.e. not in the ownership of a single individual or institution. Sheep are grazed on the fells in the summer months and brought down to lower pasture during the winter.

Fjord A deep, generally straight inlet of the sea along a glaciated coast. A fjord is a glaciated valley which has been submerged either by a post-glacial rise in sea level or a subsidence of the land.

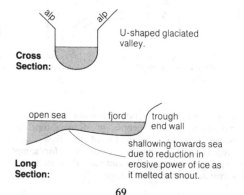

Cross Section:

alp   alp

U-shaped glaciated valley.

Long Section:

open sea   fjord   trough end wall

shallowing towards sea due to reduction in erosive power of ice as it melted at snout.

Flash flood A sudden increase in river **discharge** and overland flow due to a violent rainstorm in the upper river **basin**.

Flood plain The broad, flat valley floor of the lower course of a river, planed by annual flooding and by the lateral and downstream movement of **meanders**.

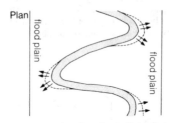

Due to concentration of erosion on the outer banks of meanders, they 'migrate' across and along the flood plain.

Flow line A diagram showing volumes of movement, e.g., of people, goods or information between places. The width of the flow line is proportional to the amount of movement, for example to portray commuter flows into an urban centre from surrounding towns and villages:

Flux For example **limestone**; used in the **blast furnace** to allow the removal of iron from the ore in the presence of heat.

Fodder crop A crop grown for animal feed, either for direct feeding, e.g. turnips, or for making **silage** as is the case with sown meadow.

Fold A buckling in formerly horizontal **strata**, caused by the process of mountain building as described below.

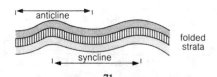

71

Fold mountains For example the Andes, Himalayas, Rockies, Alps — mountains formed when relatively soft sedimentary **strata** are compressed and uplifted between two converging **plates** of the earth's **crust**.

The sequence of events can be summarized in the following series of diagrams:

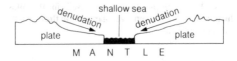

Two plates are denuded and the resulting **sediments** accumulate in the shallow sea which occupies the **geosyncline** or trough between the plates.

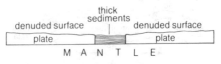

Eventually the **denudation** process leads to the accumulation of very thick sedimentary deposits.

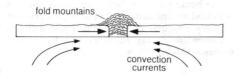

fold mountains

convection currents

Earth movements, specifically the drag exerted by **sub-crustal convection currents**, cause the plates to move towards each other. The soft sediments between are compressed and uplifted, forming fold mountains. The whole process takes many millions of years. Fold mountains are one of several features associated with **plate** boundaries.

Footloose industry Traditional **heavy industry** is restricted to specific locations, usually **resource** sites especially on coalfields. Modern **light industry** is less tied to particular locations, depending, for example, on easily available electricity for power. It is thus possible to locate close to potential purchasers to maximize market success. Such industries as publishing, assembly and electrical goods are regarded as footloose.

Fossil fuel Any naturally occurring carbon or hydrocarbon fuel, notably coal, oil, peat and natural gas. These fuels have been formed by decomposed prehistoric organisms.

The burning of fossil fuels, particularly since the **industrial revolution** has led to problems of **acid rain** in many countries of the world. Fossil fuels are nonrenewable energy resources (see **Renewable/nonrenewable resources**).

Friction of distance The intensity of human activity tends to decrease as distance from or between **nodes** increases. For example, the volume of traffic between a pair of cities decreases as they are located further apart. The amount of investment in **agriculture** decreases with increasing distance from the market town. The cost of land decreases with increasing distance from the city centre. The number of telephone calls decreases over progressively greater distance. The general model for the friction of distance is

$$f = a\frac{1}{d^2}$$

Where:

f is the volume of movement

d is distance

a is constant

See also **Gravity model**. The factors responsible for the friction of distance are cost and time; both increase with distance and act as a deterrent to movement. But improvements in, e.g., transport technology or rising incomes can reduce the friction of distance: the relative distance (though not the absolute distance) between places can therefore be decreased.

Front A boundary between two air masses. See also **Depression**.

Fulani A people of northern Nigeria practising **nomadic pastoralism**.

Functions The term used for goods and **services** available in a **central place**: in general the number and variety of functions offered increase with the size of the **settlement**. Everyday or convenience (low order) functions are available in small settlements, while these plus specialized or durable (high order) functions are available in large settlements. Small settlements may offer only two or three functions; large settlements many hundreds. See also **Central place theory**.

Gentrification The improvement of houses in **inner city** areas. This usually involves modernization of the building, e.g., putting in new windows and insulation. **Household amenities** may also be improved, e.g., an outside toilet may be transferred to the interior of the house. Gentrification can cause problems for some inner city areas as the modernized houses may be bought by wealthy outsiders. This can change the social structure of the community and the visual appearance of the **environment**.

Geosyncline The **depression**, occupied by a shallow sea, which lies between two advancing plates in **fold mountain** formation.

Geothermal energy A method of producing power from heat contained in the lower layers of the earth's **crust**. New Zealand and Iceland both use geysers and volcanic **springs** to create electricity and to heat their houses. In Britain, experiments in Southampton have successfully allowed heated water from rocks below the city to heat shops and offices in the city centre. In the long term, scientists are hoping to be able to tap heat from the granite rocks in Southwest England. Geothermal energy is an example of a renewable resource of energy (see **Renewable/nonrenewable resources**).

Glacial advance During a cooling of the **climate, ice sheets** and valley glaciers will extend over greater areas and to lower altitudes. This process is referred to as 'glacial advance'.

Glacial retreat Opposite conditions to **glacial advance**.

Glaciation A period of cold **climate** during which time **ice sheets** and **glaciers** are the dominant forces of **denudation**.

The last glaciation ended about 10,000 years ago, and much of Britain's landscape shows evidence of the work of ice (north of a line drawn approximately between the Thames and the Severn).

Glacier A body of ice occupying a valley and originating in a **corrie** or icefield. A glacier moves at a rate of several metres per day, the precise speed depending upon climatic and **topographic** conditions in the area in question. The Mer de Glace and Aletsch glaciers are present-day examples in the Alps.

Gondwanaland The southern hemisphere super-continent consisting of South America, Africa, India, Australasia and Antarctica, that split from **Pangaea** *c* 200 million years ago. Gondwanaland is part of the theory of **continental drift**.

Graded profile The long profile of a river's course after all irregularities have been removed by **erosion**. See diagram over page.

77

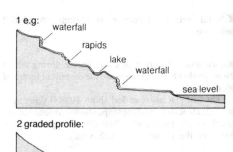

1 e.g:
waterfall
rapids
lake
waterfall
sea level

2 graded profile:
sea level

Gradient In mapwork the average gradient between two points can be calculated as:

$$\frac{\text{difference in altitude}}{\text{distance apart}}$$

Gradient between A & B:

$$\frac{300 \text{ m} - 50 \text{ m}}{800 \text{ m}}$$

e.g:

78

$$= \frac{250 \text{ m}}{800 \text{ m}} = \frac{1}{3.2}$$ | Expressed as 1:3.2 |

Interpretation: 1 metre of ascent for every 3.2 metres of horizontal equivalent.

The term gradient is also used in a more general context in **human geography** in reference to such phenomena as **population density**, land values, **settlement** ranking.

For example the urban population density gradient has the general form:

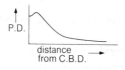

indicating a general decline in population density with increasing distance from city centres.

Granite An **igneous rock** with large crystals due to slow cooling at depth in the earth's **crust**.

Granite is a common constituent of **batholiths**, and is mainly composed of **quartz**, mica, and feld-spar minerals.

Gravity model A predictive model for forecasting volumes of movement between **nodes**. In its simplest form the gravity model is expressed:

$$Mij = \frac{PiPj}{(dij)^2}$$

Where:

M is the volume of movement between any two places i and j.

Pi, Pj are the populations of i and j.

dij is the distance between i and j.

The model suggests that the population flows, for example, between two cities will be directly proportional to the size of the cities and inversely proportional to the square of the distance between them.

The concept is borrowed from Newtonian physics. Results of gravity model predictions can be shown

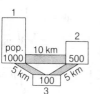

$$M_{1,2} = \frac{1000 \times 500}{10^2} = 5000$$

$$M_{1,3} = \frac{1000 \times 100}{5^2} = 4000$$

$$M_{2,3} = \frac{500 \times 100}{5^2} = 2000$$

diagrammatically in **flow line** form, as illustrated above.

In such a crude form the units of movement are not specified: the model simply portrays relative volumes of movement.

The power to which d is raised will vary according to the social and economic conditions of the area in question; low levels of development and unsophisticated transport technology will imply higher powers than 2.

The simple gravity model has been adopted in various ways, for example in Reilly's Law of Retail Gravitation (illustrated on page 82), which can be used to calculate the breaking point between spheres of influence of adjacent central places:

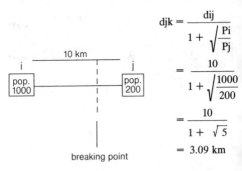

$$djk = \frac{dij}{1 + \sqrt{\dfrac{Pi}{Pj}}}$$

$$= \frac{10}{1 + \sqrt{\dfrac{1000}{200}}}$$

$$= \frac{10}{1 + \sqrt{5}}$$

$$= 3.09 \text{ km}$$

from j where djk is the distance from place j to the breaking point k.

Green belt An area of land, usually around the outskirts of a town or city, in which building and other developments are restricted by legislation. The purpose of such planning law is to attempt to preserve open space and relatively rural **environments** which would otherwise be lost in the process of **urban sprawl**.

Grid reference A method for specifying position on a map. See **Eastings**.

Groundwater Water held in the bedrock of a region, having percolated through the **soil** from the surface.

Such water is an important **resource** in areas where **surface runoff** is limited or absent.

Gryke An enlarged joint between blocks of **limestone (clints)**, especially in a **limestone pavement**. Grykes are progressively enlarged by the solution effect of rainwater.

Hade The declination of a **fault** plane from the vertical.

Hanging valley A tributary valley entering a main valley at a much higher level because of overdeepening of the main valley, esp. by glacial erosion. During glaciation, vertical erosion is much greater in the main valley than in smaller tributary valleys. Small tributary valleys will only have small glaciers in them or no glaciers at all. Once the ice has retreated the floor of the main valley lies far below the tributary valleys, which are therefore called 'hanging valleys'.

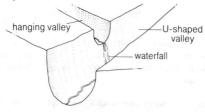

hanging valley

U-shaped valley

waterfall

Harris and Ullman model See **Multiple nuclei model**.

Headland A promontory of resistant **rock** along the coastline. See **Bay**.

Heave A horizontal displacement of **strata** at a **fault**.

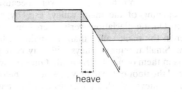

heave

Heavy industry Traditional industries using bulky **resources** are generally classified as heavy; e.g. coal mining, iron and steel-making, chemicals, engineering.

High technology approach An approach to the

development process which stresses the role of capital and sophisticated technology. It is argued that investment in large-scale **resource** management schemes (e.g. dams for **hydro-electric power**) and in the industrial sector in general is the surest way to hasten national development.

In Brazil, for example, development priorities have been identified in this way. Contrast with the **intermediate technology** (and **appropriate technology**) approach.

Hill farming A system of **agriculture** where sheep are grazed (and to a lesser extent cattle), on upland rough pasture.

In Britain hill farming occurs on **marginal land** in upland areas such as the Lake District, Snowdonia, the Scottish Highlands. The typical hill farm comprises three zones: the *inbye, intake* and **fell.** The inbye is valley-bottom land, immediately surrounding the farm buildings; it is walled or fenced and may be cultivated for **fodder crops** and sown pasture. The intake extends up the lower slopes of the surrounding **fells** and is an area of sheltered pasture for winter grazing and for lambing. The fell is an extensive area of upland rough grazing on which several farmers may have right of access.

See diagram over page.

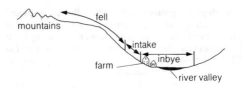

Hinterland An area inland from a port, and defined by the limit of the port's **sphere of influence**, for example:

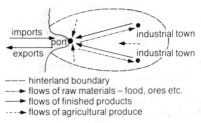

---- hinterland boundary
---·- flows of raw materials – food, ores etc.
——► flows of finished products
----► flows of agricultural produce

Histogram A graph for showing values of classed data; e.g. for a sample of data:

Income class: £ P.A.	no. of persons
1 < 4000	20
2 4000–6000	50
3 6001–8000	150
4 8001–10000	40
5 > 10000	30

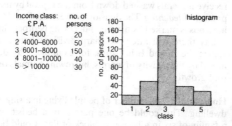

Horizon The distinct layers found in the **soil profile**; usually three horizons are identified — A, B, C as follows:

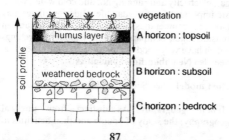

The A horizon or **topsoil** contains humus and other vegetable debris. The B horizon or subsoil will contain a larger proportion of inorganic material, and will receive minerals washed down from the topsoil by the process of *leaching*. The division between the B and C horizons is marked by a zone of decaying bedrock. In reality there will rarely be sharp divisions between zones; the A and B horizons, for example, may be mixed by the activity of worms, burrowing animals or root growth.

Household   The number of people living in a single dwelling. This could be one person in a bedsit or a family of six in a house. A block of flats would be made up of several households. Information about households is required by the ten-year national **census**.

Household amenities   Utilities in a dwelling such as gas, electricity and running hot and cold water, which are important for everyday life. Older dwellings may have poorer amenities such as shared bathroom and an outside toilet, compared to many modern dwellings which have very good amenities, e.g., central heating. See also **Neighbourhood amenities**.

Hoyt model   See **Sector model**.

Human geography   As distinct from **physical geography**; the study of people and their activities in terms of patterns and processes of population, **settlement**, economic activity and **communications**. There is no precise definition of such a broad subject, but the basic task of the human geographer is to try to explain distributions of people and their activities.

Hunter/gatherer economy A pre-agricultural phase of development in which people survive by hunting and gathering the animal and plant **resources** of **environment**. No cultivation or herding is involved. Very few hunter/gatherer societies survive today: examples include the Wiama Indians and other tribes living in the Amazon rainforest.

Hydraulic action The erosive force of water alone, as distinct from **corrasion**. A river or the sea will erode partially by the sheer force of moving water and this is termed 'hydraulic action'. For example a swift river current will undercut the outer bank of a **meander**. The force of waves breaking against a **headland** will compress the air in rock crevices and thereby cause the surrounding **rock** to shatter.

Hydro-electric power The generation of electricity by turbines driven by flowing water. Hydro-electricity is most efficiently generated in rugged **topography** where a head of water can most easily be created, or on a large river where a dam can create similar conditions. Whatever the location, the principle remains the same — that water descending via conduits from an upper storage passes through turbines and thus creates electricity.

Hydrological cycle The cycling of water through sea, land and **atmosphere** as summarized in the following diagram:

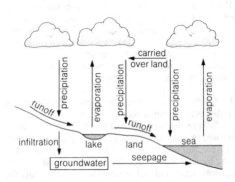

The amount of fresh water available in the system is small: 97% of all the water on the earth's surface is salt, and of the remaining 3% which is fresh, the vast majority is locked in **ice caps**.

Hygrometer An instrument for measuring the relative humidity of the **atmosphere**. It comprises two thermometers, one of which is kept moist by a wick inserted in a water reservoir. Evaporation from the wick reduces the temperature of the 'wet bulb' thermometer, and the difference between the dry and wet bulb temperatures is used to calculate relative humidity from standard tables.

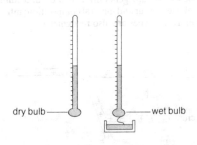

dry bulb ———   ——— wet bulb

Ice age   A period of **glaciation** in which a cooling of **climate** leads to the development of **ice sheets, ice caps** and valley glaciers. The most recent ice age was the Quaternary glaciation, ending about 10,000 years ago.

Ice cap   A covering of permanent ice over a relatively small land mass, e.g. Iceland.

Ice fall   An area of fractured ice in a **glacier** where a change of **gradient** occurs.

Ice sheet A covering of permanent ice over a substantial continental area such as Antarctica.

Igneous rock A **rock** which originated as **magma** (molten rock) at depth in or below the earth's **crust**. Igneous rocks are generally classified according to crystal size, colour and mineral composition; intrusive and extrusive types are also recognized.

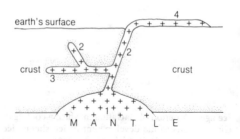

1) **Batholith**: large body of magma intruded into the earth's crust; this cools slowly at depth to form igneous rocks with large crystals such as **granite**.

2) **Dyke**: vertical or semi-vertical sheet of igneous rock; a minor intrusion compared with a batholith. Dolerite is a common dyke rock.

3) **Sill**: horizontal or semi-horizontal minor intrusion; sills and dykes exploit lines of weakness (e.g. **joints, bedding planes, faults**) in the crustal rocks.

4) **Lava flow**: extrusive igneous rocks are those which reach the earth's surface via some form of volcanic eruption. Such rocks have small crystals due to rapid cooling on the earth's surface. **Basalt** is a common example.

The terms volcanic, hypabyssal and plutonic are also used to describe igneous rocks: **volcanic rocks** are those which are extruded onto the surface; hypabyssal rocks are those of intrusions such as sills and dykes, and **plutonic rocks** are those of deep intrusions such as batholiths.

Impermeable rocks   Rocks which do not allow the passage of water, being non-porous.

Impervious rocks   Rocks which do not allow the passage of water owing to an absence of cracks or fissures. Thus an **impermeable rock** such as **granite** may be pervious due to the presence of **joints**.

Imports   Goods or services bought in from a foreign country (compare **exports**).

Inbye   See **Hill farming**.

Industrial estate   A purpose-built facility for the location of new industry, often located at the edge of an urban area where land is more readily available and where congestion is less. Motorway intersections or access points are favoured locations for industrial estates owing to the important role of road transport today.

Industrial inertia The tendency for industry to retain original locations even though such locations may no longer be optimum. Steel-making, for example, continues at Sheffield, even though the ores upon which the industry was originally based are long since worked out. The woollen **textile industry** continues to be concentrated in West Yorkshire, even though local factors such as swift streams for water power and, later, **coal** for steam power are no longer relevant. Industries retain their original locations in this way because the costs of relocating are likely to be expensive. Large amounts of capital are tied up in factory facilities; a skilled labour force may have developed in the region; economies of **agglomeration** and scale may have been established locally. For such reasons it may be cheaper to maintain the original location and to import raw materials from elsewhere. In the case of Sheffield, for example, scrap is now the major raw material for the **iron and steel industry**; the optimum location for an iron and steel works is on the coast for maximum accessibility to imported ore.

Industrialization The development of industry on an extensive scale, as happened during the Industrial

Revolution in Western Europe. Some people identify
industrialization with economic development, but for
some nations other methods of development may be
more appropriate. The term 'industrializing' is
sometimes used to describe nations mid-way through
the development process; by the same measure
'pre-industrial' and 'postindustrial' nations are
recognized.

Industrial location The optimum location for an
industry is one where the costs of transport are
minimized, with respect to both **raw materials** and
finished products. The standard theory of industrial
location is that of Alfred Weber's *locational triangle*:
an industry may locate anywhere within the triangle,
depending on the balance of transport costs relating to
the three factors.

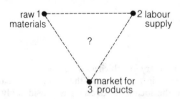

Various examples can be envisaged; e.g. an industry
with bulky raw materials which are expensive to
transport will locate close to factor 1.

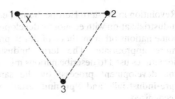

Restriction may also encourage a plant to some undesirable location or force a plant to build a new location, thereby restricting the more advantageous sites. Industries dealing with services rather than goods have to weigh the development possibilities of the sites they need to expand, but for the time being ease is [illegible]

An industry such as printing and publishing will locate close to the market, since distribution costs are the most important consideration:

An industry for which all transport costs are similar will locate in the centre of the triangle. In reality of course the locational decision is considerably more complex: factors such as the availability of power and the suitability of the land will also be taken into account, as will less tangible issues such as managers' personal preferences and outlook. Government intervention, for example through **development area** policy, may also affect the locational decision. See **iron and steel industry** for an example of shifting location over time.

Industrial Revolution The period in Britain's history when the invention and application of industrial processes led to the establishment of the world's first urban/industrial society. The period in question was approximately 1780 to 1900. The invention of the steam engine provided the key to a number of crucial industrial innovations in mining, transport and manufacture. Prior to the industrial revolution, successes within agriculture had been due to increased mechanization and efficiency, plus greater profits. This released labour to work in the industrial cities. Throughout the nineteenth century large-scale **migration** to urban areas continued, and it was during this period that the major industrial concentrations were established, for example in West Yorkshire, South Lancashire, the West Midlands, the Central Valley of Scotland.

Informal Economy Employment frequently found in **Third World** cities, characterized by irregular hours and wages, limited equipment and machinery, and often operating outside the law. Examples of informal jobs would include taxi drivers who transport materials around the city for business people, shoe shiners, and traders selling goods from trays e.g.

sweets, chewing gum, fruit. In some third World cities it is estimated that the informal sector employs between 40 to 60% of the working population.

Infrastructure The basic structure of an organization or system. The infrastructure of a city would include, e.g., its roads and railways, schools, factories, and power supplies.

Inner city The central business district (see **CBD**) and the old industrial and residential areas adjacent to it; the term inner city is usually applied in the context of the old housing of this area. Inner city housing is of two broad categories: old **terrace** or back-to-back property often in need of renovation or renewal, and new public housing in the form of high-rise flats, maisonettes and terraces.

Inorganic fraction That proportion of the **soil** which is composed of **rock** and mineral fragments; contrast with the **organic fraction**.

Intake See **Hill farming**.

Integrated steelworks An industrial site where all iron and steel manufacturing processes are carried out 'under one roof', e.g., iron manufacture, steel-making, steel rolling and processing and alloy manufacture.

Intensive farming Contrast with **extensive farming**;

intensive farming is a system of **agriculture** where relatively large amounts of capital and/or labour are invested on relatively small areas of land. An example of intensive farming is **market gardening**, where large investment in the form of greenhouses, fertilization, **irrigation** and heating systems, occurs on small holdings of land close to urban centres. In such a case yields per unit area are high. The small land holdings reflect the cost of land close to market. See **Von Thünen theory** for further details.

Interlocking spurs Typical features of the upper stages of a river's course, where vertical **corrasion** predominates over lateral corrasion and where **meanders** are in their infancy.

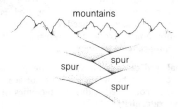

Intermediate area A category within **development area** policy; areas with problems of economic stagnation or decline of less critical proportions than those of the development areas. Under the 1978

legislation, much of Yorkshire and Lancashire, for example, became intermediate areas.

Intermediate technology Equipment and facilities of simple, cheap, practical design directly relevant to the immediate needs of a community in a developing country. The harnessing of water **resources** is an example: rivers may be controlled by complicated dams and **barrages** which may be used for **hydro-electric power** and subsequent industrial development. Such a **high technology approach** would need foreign capital, equipment and expertise. Alternatively streams and tributaries could be regulated by a series of small earth dams to provide local flood control and irrigation water. Such dams could be built with local labour and equipment, could be repaired easily and would be cheap. It is argued that such an intermediate technology approach is more relevant to the immediate needs of communities and more appropriate in conditions of scarce capital. See also **Appropriate technology**.

Internal processes Landscape-forming processes which originate below the earth's surface in the movements connected with **plate tectonics**: folding, faulting, **vulcanicity**.

Interwar suburbs Classic 'suburbia' of a typical city; the interwar suburbs comprise the extensive estates of middle-class semidetached and detached housing which mark the first wave of **urban sprawl** after 1920.

This sprawl was linked to improvements in urban transportation by train, tram and bus. The interwar suburbs lie between the Victorian **terraces** of the inner city and the commuter belt of post-Second World War housing.

Intrusion A body of igneous rock injected into the earth's **crust** from the **mantle** below. See **Igneous rocks, Dyke, Sill, Batholith**.

Iron and steel industry A key element in the heavy industrial structure of a nation; in Britain the location of the iron and steel industry is an example of changing optimum location over time: the total pattern of location has three elements. Firstly there are the coalfield locations such as Sheffield and Motherewell, dating from the early establishment of iron and steel manufacture using **coal** and blackband iron ore, found in the coal measures, as **raw materials**. Secondly there are the orefield locations such as Scunthorpe (and formerly Corby), developed in the mid-twentieth century to exploit the iron ore deposits of Jurassic rocks in eastern England. By this stage the steel-making process depended less on coal, owing to the development of the electric arc furnace and, in traditional **blast furnaces**, more efficient design which required less coal. Thirdly there are the most recently established coastal locations of the iron and steel industry, for example at Port Talbot in South Wales, and on Teesside in Northeast England.

These locations reflect the dominant role today of imported ores from places such as Scandinavia and Canada. Thus the three locational elements of the overall distribution of the iron and steel industry reflect the three stages of its evolution. Steel-making in general is in serious decline today due to a fall in demand and the development of alternative materials such as plastics: several inland steelworks have closed (e.g. Shotton, Consett, Corby).

Irrigation A system of artificially watering the land in order to grow crops. Irrigation is particularly important in areas of low or unreliable rainfall. Some of the oldest methods of irrigation were developed along the river Nile and consisted of simple devices such as the shelduf and the Archimedian Screw. Some of these methods are still used in the **Third World** today. The modern method of irrigation (often associated with countries of the developed world where more capital is available for investment in large-scale projects) is to build a dam across a river and create a reservoir. Water is then removed from the reservoir at times of need by a system of pipes and channels. In parts of Britain where rainfall may be unreliable, farmers use hose pipes and sprinklers, e.g., in the **arable farming** regions of East Anglia.

Isobar A line joining points of equal atmospheric pressure, as on the meteorological map on the opposite page.

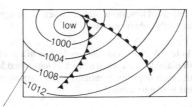

Isobar, indicating atmospheric pressure in millibars.

Isohyet A line on a meteorological map joining places of equal rainfall.

Isotherm A line on a meteorological map joining places of equal temperature.

Joint A vertical or semi-vertical fissure in a **sedimentary rock**, contrasted with roughly horizontal **bedding planes**. In **igneous rocks** jointing may occur as a result of contraction on cooling from the molten state. Joints should be distinguished from **faults** in that they are on a much smaller scale and there is no relative displacement of the rocks on either side of the joint. Joints, being lines of weakness, are exploited by **weathering**.

Karst topography Named after a region in Yugoslavia, the term 'karst' is used more broadly to

103

describe areas of **limestone** scenery where **drainage** is predominantly subterranean.

Kettle hole A small depression or hollow in a glacial **outwash** plain, formed when a block of ice embedded in the outwash deposits eventually melts, causing the **sediment** above to subside.

Labour intensive A system of **agriculture** or industry where labour (rather than capital) forms the major input. For example, rice cultivation in much of Southeast Asia is a labour-intensive activity with very little mechanization. As nations develop, capital tends to replace labour in certain sectors of the economy and labour intensity decreases.

Laccolith An igneous intrusion, domed and often of considerable dimensions, caused where a body of viscous **magma** has been intruded into the **strata** of the earth's **crust.** These strata are buckled upwards over the laccolith.

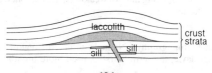

Lagoon An area of sheltered water behind a **bay bar** or **tombolo**.

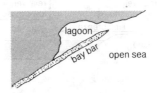

The calm water behind a coral reef is also referred to as a lagoon.

Land breeze A breeze occurring at night when the sea is relatively warmer than the land and when, as a result, pressure is relatively lower over the sea. During the day opposite conditions prevail; the land is relatively warmer and the pressure gradient is from land to sea; a **sea breeze** then occurs. Such conditions arise because the land heats up and cools down more quickly than the sea. See diagram over page.

Land reform The process of redistributing land, especially in developing nations where current circumstances may be against fair access to land and against agricultural improvement. For example, in parts of Latin America, the traditional **land tenure** system is made up of large commercial estates (known

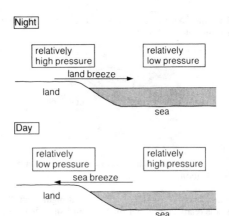

as *estancias*) and small peasant holdings. Often the estate occupies the best farming land whilst the peasants farm in difficult conditions such as on valley sides. This inequality has led to pressure for land reform, sometimes by violent revolution. The aim of recent land-reform schemes has been to provide typical farming families with security, reasonable farming land, and an incentive to improve productivity. Concern for land reform occurs not only in situations of unequal distribution as described above, but also in traditional rural economies where land is allocated through the chieftainship, often in small and fragmented parcels, as occurs in parts of Africa. Here the necessary reforms would be consolidation of holdings, the provision of secure tenure, and some facility for the progressive farmer to improve productivity. However, clumsy land reform may do more harm than good: land in many rural societies is much more than a commodity, it is an integral part of culture and heritage.

Land tenure A system of land ownership or allocation.

Land-value gradient 'Average land values per unit area decline with increasing distance from the central business district'. Whilst this is usually true, some suburban areas with good **accessibility** may prove to be good locations for business and retail centres. Therefore the land-value gradient may show peaks within suburban areas.

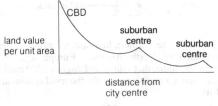

Laterite A hard (literally 'brick-like') soil in tropical regions caused by the baking of the upper **horizons** by exposure to the sun. Laterite occurs often as a result of mismanagement of the tropical **environment**, for example by clear-felling of **tropical rain forest**, or by a decrease in the **regeneration** cycle in **shifting agriculture**.

Laterite has low agricultural potential; it is difficult to cultivate with simple tools and has a low nutrient content owing to the removal of forest and consequent reduction in nutrient cycling.

Latitude Distance north or south of the equator, as measured by degrees of an angle measured from the earth's centre:

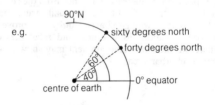

Laurasia The northern hemisphere supercontinent consisting of North America, Europe and Asia excluding India, that split from **Pangaea** *c* 200 million years ago. Laurasia is part of the theory of **continental drift**.

Lava **Magma** extruded onto the earth's surface via some form of volcanic eruption. Lava varies in **viscosity**, colour and chemical composition. Acidic lavas tend to be viscous and flow slowly, basic lavas tend to be non-viscous and flow quickly. Commonly **lava flows** comprise basaltic material, as for example in the process of **sea-floor spreading**.

Lava flow A stream of **lava** issuing from some form of volcanic eruption.

Lava plateau A relatively flat upland composed of layer upon layer of approximately horizontally bedded lavas, e.g. the Deccan Plateau of India.

Least cost location The optimum location for an industry in consideration of transport and production costs. The least cost location may change over time with transport developments and changing markets. The least cost location for iron and steel manufacture, for example, has shifted from inland coalfields to coastal sites as imported **raw materials** have increased in importance.

Levée The bank of a river, raised above the general level of the **flood plain** by **sediment** deposition during flooding. When the river bursts its banks, relatively coarse sediment will be deposited first, and recurrent flooding will build up the river's banks accordingly. See diagram over page.

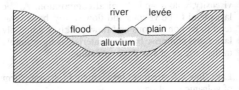

cross section of lower stage of river's course

river   levée

flood   plain

alluvium

Note that continuous deposition on river bed raises the entire channel above the general level of the flood plain.

Light industry As contrasted with **heavy industry**: light industry includes industries such as the manufacture of electrical goods, printing and publishing, distribution trades and clothing manufacture.

Light industry is generally to be found on purpose-built estates, often on the edge of an urban area where congestion is less and **communications** are good. Light industry is generally cleaner and less pollutive than heavy industry and is organized mainly for the manufacture of consumer durables.

Lignite A soft form of **coal**; harder than **peat** but softer than **bituminous coal**.

Limb One element of a **fold**, with respect to either an **anticline** or a **syncline**.

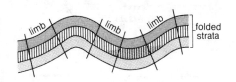

Limestone Calcium-rich **sedimentary rock** formed by the accumulation of skeletal matter of marine organisms. The most favourable conditions for limestone formation comprise a warm, clear, shallow sea.

Many limestones contain fossils: shells and skeletal traces are common. Being both soluble and **permeable**, limestone gives rise to a dry surface landscape with characteristic features of underground **drainage**, as found in the Malham district of Yorkshire, where the Carboniferous limestone, approximately 350 million years old, is the major element of the landscape.

Typical limestone features are summarized in the diagram over page.

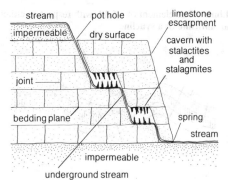

- stream
- pot hole
- limestone escarpment
- impermeable
- dry surface
- cavern with stalactites and stalagmites
- joint
- bedding plane
- spring
- stream
- impermeable
- underground stream

Limestone pavement An exposed **limestone** surface on which the **joints** have been enlarged by the solution effect of rainwater (weak carbonic acid). These enlarged joints or **grykes** separate roughly rectangular blocks of limestone called **clints**.

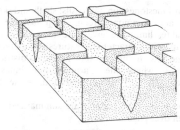

112

Link An element of a route **network**.

Load The **sediment** transported by the agents of **erosion** — rivers; moving ice; the sea. The size and volume of load transported depends upon the power of the transporting medium. In a river system, for example, more load is carried in times of high discharge. A river's load comprises material rolled or bounced along the bed, material carried in suspension, and material carried in solution. The finest sediment is carried the greatest distances and may contribute to the formation of, for example, a **delta**.

Loess A very fine **silt** deposit, often of considerable thickness, transported by the wind prior to **deposition**. In northern China, for example, there are extensive loess deposits, carried by wind from the arid plateau lands of Central Asia. Loess is extemely porous and the surface is consequently dry. When irrigated, loess, which has high inherent fertility, can **yield** high agricultural productivity.

Location A basic theme in **human geography**; the position of population, **settlement**, economic activity in space.

Location quotient A measure of the concentration of a phenomenon in space. The regional characteristics of the phenomenon are compared with its national characteristics in order to quantify the extent of regional concentration. Examine, for example, the level of employment in metal manufacture in South Wales and in Britain as a whole:

	South Wales	Great Britain
Employment in metal manufacture (thousands)	62.7	518
Total employment (thousands)	682.1	22,182

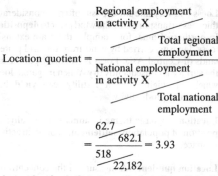

$$\text{Location quotient} = \cfrac{\dfrac{\text{Regional employment in activity X}}{\text{Total regional employment}}}{\dfrac{\text{National employment in activity X}}{\text{Total national employment}}}$$

$$= \cfrac{\dfrac{62.7}{682.1}}{\dfrac{518}{22,182}} = 3.93$$

Any value for L.Q.>1 means that the activity in question is concentrated in the region under study.

The concentration increases as the value increases..

Location triangle See **Industrial location**.

Longitude A measure of distance on the earth's surface east or west of the Greenwich Meridian, an imaginary line running from pole to pole through Greenwich in London. Longitude, like **latitude**, is measured in degrees of an angle taken from the centre of the earth. Lines of longitude are examples of *Great Circles*; an example of the latter is any circle drawn on the earth's surface, the centre of which coincides with the centre of the earth. The Great Circle route between two points on the earth's surface is the shortest possible route. The only line of latitude which is a Great Circle is the equator. The precise location of a place can be given by a **grid reference** comprising longitude and latitude: e.g. Lagos, Nigeria:

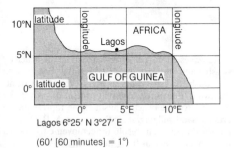

Lagos 6°25′ N 3°27′ E

(60′ [60 minutes] = 1°)

115

Longshore drift The net movement of material along a beach due to the oblique approach of waves to the shore. Beach deposits move in a zig-zag fashion as shown in the diagram below. Longshore drift is especially active on long, straight coastlines. The movement of beach material can be halted by the construction of **breakwaters**.

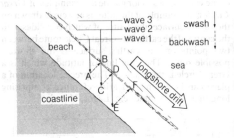

As waves approach, sand is carried up the beach by swash, and retreats back down the beach with backwash. Swash occurs at 90° to the wave front; backwash (due purely to gravity) occurs directly down the slope of the beach. Thus a single representative grain of sand will migrate in the pattern A, B, C, D, E, F in the diagram. The resulting net movement of beach material is known as 'longshore drift'.

Lopolith A basin-shaped *igneous intrusion.*

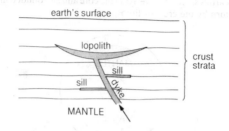

Magma Molten rock; originating in the earth's **mantle** and the source of all **igneous rocks**.

Malnutrition The condition of being poorly nourished, as contrasted with under-nourishment, which is lack of sufficient quantity of food. The diet of a malnourished person may be high in starchy foods but is invariably low in protein and essential minerals and vitamins. Malnutrition occurs in many parts of Africa and Asia and gives rise to numerous ailments which lead to high infant mortality: kwashiorkor, marasmus, beriberi and pellagra are examples. The remedy for almost all such diseases is simple yet tragically unavailable to a large proportion of the world's population — high protein foods (such as fish, meat, pulses) and green vegetables to provide minerals and vitamins.

Mantle The largest of the concentric zones of the earth's structure, overlying the core and surrounded in turn by the crust or lithosphere.

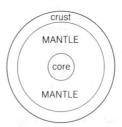

Marble A whitish, crystalline **metamorphic rock** produced when **limestone** is subjected to great heat or pressure (or both) during earth movements.

Marginal land In **agriculture** the term is used to describe land which is only just worth managing. The farmer will make only a small profit, if any, in the use of such land. Land may be regarded as marginal in either a physical or an economic sense: for example, upland areas with thin soils and harsh climates may be considered marginal, as may remote land at great distance from (or inaccessible to) the market. Much of the rough upland pasture which is an element of British **hill farming** is marginal, and farmers stay in business with the aid of government subsidies.

Maritime Climate   A **climate** that is affected by its closeness to the sea, giving a warm summer and a mild winter. Britain is an example of a country with a maritime climate, and has a small annual range of temperature. Compare **Extreme climate**.

Market area   The area from which consumers will travel into a **central place** in order to obtain goods and **services**. The size of the market area will be determined by the position of the central place in the **settlement hierarchy**: the market area (or **sphere of influence**) for low order central places is small, since individual travel tolerance for everyday goods and services is low. Conversely the market area for high order central places is large, since individual travel tolerance for specialized goods and services is high. See **Central place theory**.

Market gardening   An intensive type of **agriculture**, traditionally located on the margins of urban areas to market fresh produce on a daily basis to the city population. Typical market garden produce includes salad crops such as tomatoes, lettuce, cucumber, etc., cut flowers, fruit and some green vegetables. The classic market garden is characterized by a small landholding, high capital investment in the form of equipment such as greenhouses and fertilizers, plus high yields: these conditions are in response to the high value of land near urban areas and the perishability of the produce (see **Von Thünen theory**).

However, modern developments in **communications** and transport technology, and indeed in food processing and consumer tastes, have led to a relaxation of the traditional locational factors behind market gardening. The production of, e.g., peas, beans and cauliflowers in the English fens, primarily for canning and freezing, is an example of the 'new' market gardening; some produce is marketed direct to London via the excellent communications of eastern lowland England.

Masai Nomadic pastoralists of East Africa.

Maximum and minimum thermometer An instrument for recording the highest and lowest temperatures over a 24-hour period, the readings usually being taken at 0800 hrs.

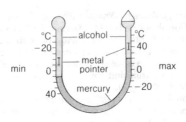

As the temperature rises, the alcohol in the left-hand tube expands, pushing the mercury up in the

120

right-hand tube, in which the alcohol also heats up and vapourizes into the cavity in the conical bulb. When the temperature falls this vapour liquefies and the alcohol in the left-hand tube contracts, causing the mercury to flow in the reverse direction. Metal pointers mark the extreme positions of the mercury column; these are reset with a magnet.

Meander A large bend, especially in the middle or lower stages of a river's course. A meander is the result of lateral **corrasion** which becomes dominant over vertical corrasion as the **gradient** of the river's course decreases. The characteristic features of a meander are summarized thus:

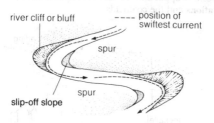

river cliff or bluff

---- position of swiftest current

spur

slip-off slope

spur

Concentration of **erosion** on the outer bank causes undercutting and the development of a **river cliff**. Slack water on inner bank causes **deposition** and a bank of **sediment** called a **slip-off slope** is produced.

In their fullest extent meanders may appear like this:

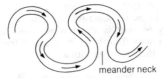

meander neck

See also **Oxbow lake**.

Merry-go-round A system of freight transport designed for continuous and largely automated collection and delivery. The concept has been successfully applied to the link between **thermal power stations** and the coalfields:

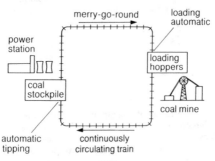

122

Mesa A flat-topped isolated hill in arid regions, for example in Arizona and New Mexico in the USA.

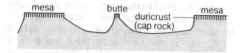

A **butte** is a relatively small outlier of a mesa. The shape of a mesa is partly determined by the occurrence of a horizontal resistant cap rock, underlain by softer, more readily eroded **sediments.**

Metamorphic rock A **rock** which has been changed by intensive heat or pressure. Metamorphism implies an increase in hardness and resistance to **erosion.** Shale, for example, may be metamorphosed by pressure into **slate; sandstone** by heat into **quartzite, limestone** into **marble.** Metamorphism of pre-existing rocks is associated with the processes of **folding, faulting** and **vulcanicity.**

Migration A permanent or semi-permanent change of residence. **Population migration** may occur at a variety of scales and for a number of reasons: international, regional and local migrations occur and their precise reason for movement will differ between individuals. Employment-regulated migration, such as is occurring in the current **Third World**

urbanization process, is dominant, but people will also move for family reasons, for retirement, or as a result of political pressure (forced migration). Scale contrasts are considerable: at the international scale movement from the British Isles to North America in the nineteenth century involved many thousands of individuals; at the local scale there has been a tendency for people to move from the inner to the outer city and even out of the city altogether. The decision to migrate, whatever the scale, is determined by the balance of *push factors, pull factors* and *intervening obstacles*. Push factors are negative aspects of life at the current place of residence; pull factors attractive aspects of some potential future location. Intervening obstacles might include such factors as distance, cost, physical and political barriers — the greater the magnitude of the obstacle the less likely the migration will be. Invariably there is *distance-decay* in migration; i.e. the number of migrants into a given destination declines with increasing distance of origins.

Mixed farming A system of **agriculture** which comprises a variety of arable and pastoral elements.

Monoclinal fold A **fold** in which one **limb** is vertical or practically so.

monoclinal fold

Monoculture The growing of a single crop. Traditional viticulture in the south of France is an example. Monoculture can be unsound in two ways: firstly the **soil** is progressively drained of specific nutrients, and secondly dependence on a single crop may be dangerous in the market if the crop fails or if consumer tastes change. Monoculture may also cause an increase in the occurrence of plant-specific pests and diseases.

Monsoon The term strictly means 'seasonal wind' and is used generally to describe a situation where there is a reversal of wind direction from one season to another. This is especially the case in Southeast Asia, where two monsoon winds occur, both related to the extreme pressure gradients created by the large land mass of the Asian continent. In summer (April to September) the intense heating of the land leads to the development of low pressure over Northwest India, and southwesterly winds are drawn in over the Indian ocean. This southwest monsoon brings heavy rain to India and Southwest Asia in general, and especially to those areas where the **orographic** effect is operating. In winter (October to March), the chilling of the Asian interior leads to high pressure and the establishment of the northeast monsoon; cold dry winds which bring little rain. In northern China they may also carry dust from the deserts of the continental interior.

Moraine A collective term for debris deposited on or

by **glaciers** and ice bodies in general. Moraine differs from fluvial **sediment** in being unsorted and composed of angular fragments. Several types of moraine are recognized: *lateral* moraine forms along the edges of a valley glacier where debris eroded from the valley sides, or weathered from the slopes above the glacier, collects. *Medial* moraine forms where two lateral moraines meet at a glacier junction; *englacial* moraine is material which is trapped within the body of the glacier; *ground* moraine is material eroded from the floor of the valley and used by the glacier as an abrasive tool. A **terminal** moraine is material bulldozed by the glacier during its advance and deposited at its maximum down-valley extent. *Recessional* moraines may be deposited at standstills during a period of general glacial retreat.

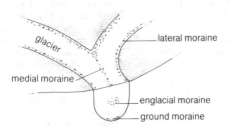

Mortlake See **Oxbow lake.**

Multinational company A company that has branches in many countries of the world, and often controls the production of the primary product and the scale of the finished article. For example, multinationals own many tea plantations in the Third World, where the tea is picked and processed by local labour (see **plantation agriculture**). The multinationals also control the selling of the tea, most of which is sent to developed countries. It is argued by some economists that labour in the Third World is badly paid in order to provide cheap products worldwide. The multinationals, however, claim to provide higher wages and better working conditions than indigenous industries.

Multiple nuclei model (of Harris and Ullman, 1945) A model of urban structure stating that most large cities develop around a number of separate centres or nuclei, rather than round a single centre (compare **Burgess model** and **sector model**). Different land uses are therefore situated around the city creating a cellular structure. The pattern of these cells or nuclei will reflect the unique factors of the site and/or history of any particular city. See diagram over page.

Multiple purpose/resource management A strategy for resource management which attempts to provide

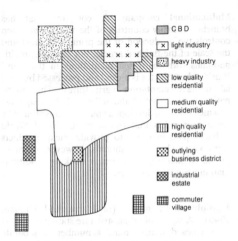

▨	CBD
⊠	light industry
▦	heavy industry
▨	low quality residential
☐	medium quality residential
▥	high quality residential
▨	outlying business district
▨	industrial estate
▦	commuter village

maximum availability of a **resource** to as wide as possible a variety of users, without endangering the quantity or quality of the resource for any particular consumer. The management of water resources is a case in point: multiple purpose dams can cater for recreation, **hydro-electric power** generation, flood control and fishing (see **Conflicting and compatible demands**). In practice, multiple purpose resource management may be an elusive goal due to the problems of accommodating widely different demands.

Myrdal model Swedish economist Gunnar Myrdal developed (1957) the concept of *cumulative causation* as a contribution to the understanding of regional development processes. Myrdal's thesis was that the adoption of innovations and progressive ideas were most likely to occur in areas where economic and social development was already established; i.e. in urban centres, especially the national core (see also **core/periphery** model of Friedman). Thus there would be self-perpetuating growth and development in major urban centres, as summarized on the diagram over page.

National park An area of scenic countryside protected by law from uncontrolled development. A national park has two main functions: (a) to conserve the natural beauty of the landscape; (b) to enable the public to visit and enjoy the countryside for leisure and recreation.

Natural arch A feature of coastal **erosion**, resulting from the differential erosion of hard and soft **rocks**. For example softer rocks in a **headland** may be rapidly eroded, first to produce two caves and later to create a natural arch as the backs of the two caves are eroded

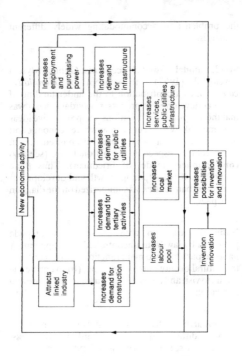

through. Eventually the roof of the arch may collapse to produce a **stack**.

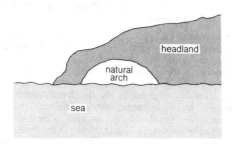

Natural increase The increment in population due to the difference between **birth rate** and **death rate**: in the **demographic transition**, for example, the phase of rapid population increase is the result of a constant high birth rate and a rapidly falling death rate. Natural increase is one component of *gross* population increase: in-migration can also cause a rise in total population. The complete **population change** equation for a nation or region will thus comprise both natural and migrational factors.

Nearest neighbour analysis See **Spatial distribution**.

Neighbourhood amenities Useful facilities in the

local area. Many neighbourhood amenities will be provided by the local council, e.g., swimming pools, park benches, bus shelters and street lights. See also **Household amenities**.

Neocolonialism A maintenance of the authority/dependency relationship of the colonial period through economic mechanisms. Although the majority of former colonies, of Britain for example, have gained political independence, they remain economically dependent on the former colonial power. Many industrial nations maintain close connections with their former colonies, often through multinational companies and trading links. Products such as tea, coffee, tin and copper have their prices fixed in the industrial nations and not in the producing nations. Thus although the period of political colonialism is largely over, economic colonialism persists.

Network A system of **links** and **nodes** via which flows of communication are passed. The structure of networks is represented diagrammatically as follows:

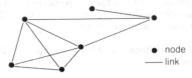

See also **Accessibility matrix, ß index, Connectivity.**

Névé Compact snow. In a **corrie** icefield, for example, four layers are recognised: blue and white ice at the bottom of the ice mass; névé overlying the ice, and powder snow on the surface.

New town A new urban location created for one of two purposes, or both: (a) to provide overspill accommodation for a large city or **conurbation**; (b) to provide a new focus for industrial development, for example in a depressed region. In the immediate post-war period a number of new towns were designated in Southeast England: for example, Stevenage, Harlow, Basildon — all designed to provide overspill accommodation for the population of London. More recently towns such as Telford and Newtown (designated in 1963 and 1967 respectively) have been created as foci for regional development.

Nivation A type of physical **weathering** whereby **rocks** are denuded by the freezing of water in cracks and crevices on the rock face. Water expands on freezing into ice and this process causes stress and fracture along any line of weakness in the rock. Nivation debris accumulates at the bottom of a rock face as **scree**. Nivation is particularly active in cold **climates** or at altitude, for example, on the exposed slopes above a **glacier**.

Node A **central place**; in **network** studies **settlements** are referred to as nodes, and more generally the term implies an urban centre which is nodal to a surrounding region.

Nomadic pastoralism A system of **agriculture** in dry grassland regions: examples of societies which are based on nomadic pastoralism are the Masai of East Africa and the Fulani of northern Nigeria. People and stock (cattle, sheep, goats) are continually moving in search of pasture and water: the low rainfall of much of the African **savannas** and the scanty nature of the grazing vegetation require a nomadic cycle of activity. The pastoralists subsist on meat, milk and other animal products.

Nomadic pastoralism is under pressure from a variety of causes: severe drought in the Sahel region (the southern margins of the Sahara desert) in the late 1970s led to destitution among thousands of wandering herdspeople. Overgrazing and the trampling of the soil surface around boreholes has hastened the process of **desertification**.

Various schemes, sponsored, for example, by the United Nations, have been introduced in an attempt to improve conditions for the pastoralists: for example the introduction of high quality stock, parasite control and pasture improvement. The chief problem is that most such schemes require a fundamental change in the nomad's traditional life-style: permanent **settlement** is an integral part of most development schemes.

Northings An element of a **grid reference**. See **Eastings**.

Nuclear power station An electricity-generating plant using nuclear fuel as an alternative to the conventional thermal fuels of **coal** and oil. Nuclear power stations, while expensive to construct, are relatively cheap to run but have to be built in remote (often coastal) locations well away from population concentrations. This is partly in response to public anxiety over the safety of such stations and due to the problems of radioactive waste disposal.

Nunatak A mountain peak projecting above the general level of the ice near the edge of an **ice sheet**. Such features occur, for example, in Greenland and Antarctica.

Nutrient cycle The cycling of nutrients through the **environment** as, for example, in the process of leaf-fall in a woodland **ecosystem** whereby fallen vegetation is broken down by bacteria into constituent nutrients such as calcium and magnesium: these are released into the **soil** where they are available for root uptake and renewed growth of vegetation. And so the cycle continues on an annual basis. The actual pathways of nutrients through an ecosystem are very complex, but all involve the use of nutrients by living organisms for growth, and the subsequent release of nutrients back to the environment on death and decay.

Offshore bar A low bank of sand and shingle lying some distance offshore and exposed at high tide: the offshore bar is created when a very gently shelving seabed causes waves to break well away from the actual shoreline. The Cape Hatteras coastline of the Atlantic coast of the USA is an area where such conditions prevail.

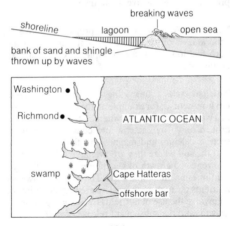

Open cast A type of mining where the mineral is extracted by direct excavation rather than by shaft or drift methods. For example in parts of the Yorkshire coalfield the coal measures occur very close to the surface and the superficial overburden is relatively easily removed. The **coal** is then excavated by mechanical grabs and removed by trucks.

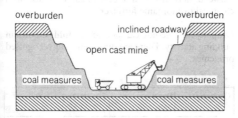

Such mining creates extensive scars in the landscape which, if left unmanaged, represent serious environmental deterioration. The National Coal Board is required to undertake *landscaping* after the cessation of mining: this usually involves the infilling of the open cast site and the planting of vegetation on the reclaimed surface. The vegetation will also help to stabilize the infil.

Organic fraction That proportion of the **soil** which

is composed of material derived from the breakdown of vegetation or other organic matter. This is to be contrasted with the inorganic fraction which is the particulate material derived from the **weathering** of bedrock.

Organic fertilizer A fertilizer composed of organic material, e.g. horse manure, farmyard manure, seaweed derivatives and bonemeal. Contrast with chemical or inorganic fertilizer.

Orogeny A geological period of **fold mountain** building activity. There are three generally recognised orogenies:

Orogeny	Peak date	Examples
Caledonian	Approx 400 million years ago	Scottish Highlands
Hercynian	Approx 300 million years ago	Appalachians, USA
Alpine or Tertiary	Approx 50 million years ago	Rockies, USA Andes, South America Alps, Europe Himalayas, Asia

Orographic rainfall (or relief rainfall) Precipitation caused by the rising of air over, for example, a coastal mountain range:

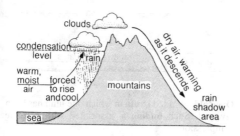

The area inland from the coastal range is likely to experience very low rainfall since the air has lost its moisture on the seaward flank of the mountains. Such a situation obtains on the western seaboard of North America where the Rocky Mountains generate the orographic effect.

Outwash Sedimentary material deposited by meltwater issuing from a **glacier** snout or **ice sheet** margin.

Outwash sands and gravels differ from glacial **moraines** in being *sorted* and subjected to **attrition.** The coarsest **sediments** are deposited close to the ice margin; fine material will be transported greater distances before **deposition**.

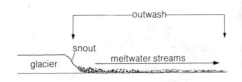

Overfold Folded **strata** in which one limb has been inverted. Such **folding** is associated with extreme compression.

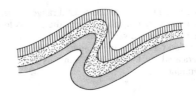

Oxbow lake (or **mortlake**, or **cut-off**) A severed **meander** in the middle or lower stages of a river's course. An oxbow lake is formed when the neck of a meander is breached by **erosion** on the outer banks. See diagram over page.

140

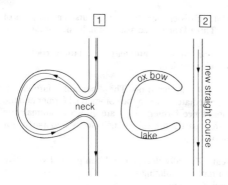

In time the oxbow lake will silt up; pioneer vegetation will invade and the lake will eventually disappear.

Pangaea The supercontinent or universal land mass in which all continents were joined together approximately 200 million years ago. The theory of Pangaea's existence was devised by Alfred Wegener in 1912. See **Continental drift**.

Pastoral farming A system of farming in which the raising of livestock is the dominant element. In the commercial context this may refer to, for example, dairy farming in Britain or sheep rearing in Australia, while subsistence pastoralism occurs in many parts of the **Third World**. See also **Nomadic pastoralism**.

Pearson product moment correlation coefficient See **Correlation**.

Peasant agriculture The peasant farmer is intermediate between the commercial farmer and the subsistence farmer: crops are grown or animals are reared partly for subsistence needs and partly for market sale.

Peat Partially decayed and compressed vegetative matter accumulating in areas of high rainfall and/or poor **drainage**.

In Britain, peat occurs in the upland areas of the north and west, forming *blanket-bog* over much of the Pennines, and in low-lying parts of East Anglia. Peat **soil** results from the limited breakdown of fallen vegetation in *anaerobic* conditions: total breakdown into humus cannot occur in waterlogged, airless conditions. Upland peat is acidic and infertile due to leaching; lowland peat has a higher nutrient status and is more useful for **agriculture**.

Pedestal rock See **Zeugen**.

Peneplain A region that has been eroded until it is almost level. The more resistant rocks will stand above the general level of the land. See **Cycle of erosion**.

Per capita income The mean annual per capita income in a nation or region is the average income per head, per year; i.e. the average income of an individual wage-earner in the population in one year. Per capita income comparisons are used as one indicator of levels of economic development.

Periglacial features Landscape characteristics resulting from proximity to glacial conditions. A periglacial landscape is one which has not been glaciated *per se*, but which is affected by the severe **climate** prevailing around the ice magin. Intensive **nivation** is characteristic of periglacal **environments**, as is *solifluction,* a process whereby thawed surface soil creeps downslope over a permanently frozen **subsoil** (permafrost). Much of the Canadian **tundra** and Siberian heartland are affected by such periglaciation.

Periphery A remote and/or underprivileged region as in the core/periphery model (see **Core**). Such regions are generally lacking in resources and offer little development opportunity, and as such are the last to be integrated into the national development process.

Permeable rock A **rock** that is permeable and which will allow water to pass through freely: such rocks are usually porous.

Pervious rocks, whilst non-porous, will allow water to pass through via **joints** and fissures.

pH A measure of soil acidity/alkalinity. A pH value of 7.0 is regarded as neutral, while pH values of less than 7.0 indicate acidic conditions. pH values greater than 7.0 indicate increasingly alkaline conditions. The optimum conditions for cereal growth are indicated by a value of about 6.5.

Physical geography The study of our **environment**, comprising such elements as geomorphology, hydrology, pedology, meteorology, climatology and biogeography.

Pie chart A circular graph for displaying values as proportions:

(Imaginary data)

Journeys to work
(sample of urban population)

Mode	No.	%	Sector° (% × 3.6)
Foot	25	3.2	11.5
Cycle	10	1.3	4.7
Bus	86	11.2	40.3
Train	123	15.9	57.2
Car	530	68.5	246.3
Total	774	100	360
		percent	degrees

144

Plantation Agriculture A system of **agriculture** located in a tropical or semi-tropical **environment**, producing commodities for export to Europe, North America and other industrialized regions. Coffee, tea, bananas, rubber, sisal are examples of plantation crops.

Plantation agriculture is distinctive in that it is a form of **commercial agriculture** located in a generally subsistence or peasant environment: it is an extension of the commercial agriculture of the developed world into a generally **Third World** environment. Some plantations are run and financed by **multinational companies** and the profits from such operations are generally channelled back to Europe or North America. As a result many plantations are institutions of **neocolonialism**. There is a worry that plantations often take up valuable farmland, growing commodities required by the richer countries. Thus more valuable local crops are forced onto poorer land. In areas where unemployment is high, plantations have traditionally paid low wages. On a more positive side, many plantation operators provide such facilities as housing, education and health care for their workers, as well as a plot of land. But it is difficult to avoid the conclusion that plantation agriculture in its traditional form is unacceptable in view of contemporary development priorities in Third World nations.

Plate tectonics The branch of geology which deals

with the processes related to the movement of the sections of the earth's **crust**, termed *plates*. The crust is composed of a number of plates of varying dimensions, which move under the influence of sub-crustal convection currents in the earth's **mantle**. Such movement accounts for continental drift, whereby the present-day continents are regarded as diverging fragments of a former supercontinent known as **Pangaea**.

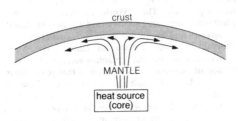

Such processes as faulting, folding, earthquakes and **vulcanicity** are all resultant upon the movement of plates: they occur primarily at *plate boundaries* which may be classified as *diverging* or *converging*. At divergent plate boundaries, new basaltic crustal material is constantly being formed by the upwelling of **magma**. This process is called *seafloor spreading* and occurs, for example, along the length of the Mid-Atlantic Ridge. See diagram over page.

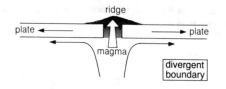

At convergent boundaries, the sinking plate may melt and lead to volcanic activity as follows:

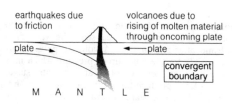

The zone of earthquakes and volcanic activity is called the *subduction zone*. See also **Fold mountains**.

Plucking A process of glacial **erosion** whereby, during the passage of a valley glacier or other ice body, ice forming in cracks and fissures drags out material from a **rock** face. This is particularly the case with the backwall of a **corrie**.

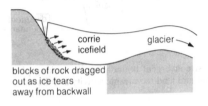

corrie icefield

glacier →

blocks of rock dragged out as ice tears away from backwall

Plug The term applied in studies of **vulcanicity** to the solidified material which seals the vent of a **volcano** after an eruption. A volcanic plug is thus responsible for the build-up of pressure which may result in an explosive eruption at some later stage. Viscous **lavas** produce the most effective plugs and, because of their resistance to **erosion**, volcanic plugs tend to stand out in the landscape when softer surrounding material has been worn away.

Plunge pool See **Waterfall**.

Plutonic rock **Igneous rock** formed at depth in the earth's **crust**: crystals are large due to the slow rate of cooling, and **granite**, such as is found in **batholiths** and other deep-seated intrusions, is a common example.

Podzol Characteristic **soil** of the **taiga** coniferous forests of Canada and the northern USSR: podzols are leached, greyish soils: iron and lime expecially are leached out of the upper horizons, to be deposited as *hardpan* in the B **horizon**.

Pollution Environmental damage caused by improper management of **resources**, or by careless human activity.

Great progress has been made in the control of some of the worst causes of pollution, such as by persistent agricultural chemicals and by heavy metals, but other forms give equal and growing cause for concern. Smoke control legislation, for example, has led to a vast reduction in pollution of the **atmoshphere**, and as a result cities are cleaner and the number of people suffering from respiratory diseases has dropped. However, industrial emissions remain highly pollutive: much of the sulphur dioxide emitted by British industry is blown by the prevailing winds over Scandinavia, where the resulting **acid rain** has led to the poisoning of entire **ecosystems**: many lakes are now devoid of fish; forest growth is stunted, and public water supplies require calcification. Pollution can take many forms: aircraft noise is a form of pollution, as is the unsightliness of a rubbish dump.

Population change The increase or decrease of a

population, the components of which are summarized as follows:

Flow diagram for the population change system:

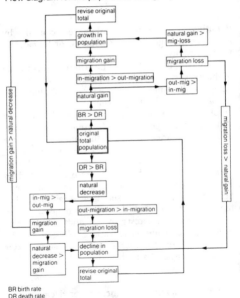

BR birth rate
DR death rate

Population density The number of people per unit area. Population densities are usually expressed per square kilometre, and can range from less than one in remote, inhospitable regions, to many hundreds in urban areas or on highly productive agricultural land.

Population distribution The pattern of population location at a given **scale**. At the global scale, population distribution shows a concentration in specific areas, for example in parts of Asia and Europe, and is sparse in others, such as in the polar regions and the hot deserts.

Population explosion On a global **scale**, the dramatic increase in population during the twentieth century. The graph of world **population growth** is as follows:

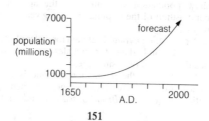

The first thousand million was reached by about 1820; the second by about 1930; the third by 1960 and the fourth by about 1975. The current world population is of the order of 4000 million and is expected to double by the end of the century.

The major cause behind the increasing growth rate is a world-wide fall in **death rate**, resulting from a complex of development factors such as improved nutrition and health care and better **communications**. See **Demographic transition**.

Population growth An increase in population of a given region: this may be the result of natural increase (more of births than deaths) or of in-migration, or both.

Population migration See **Migration**.

Population pyramid A type of **bar graph** used to show population structure, i.e., the age and sex composition of the population for a given region or nation.

A population pyramid has data for males plotted on one side and data for females on the other. See diagram on opposite page.

The shape of the population pyramid is a useful indicator of the stage of development reached by the nation in question. *Broad-based* pyramids indicate

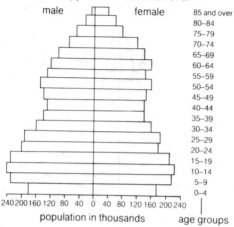

Population pyramid for Scotland 1976.

male	female	age groups
		85 and over
		80–84
		75–79
		70–74
		65–69
		60–64
		55–59
		50–54
		45–49
		40–44
		35–39
		30–34
		25–29
		20–24
		15–19
		10–14
		5–9
		0–4

240 200 160 120 80 40 0 40 80 120 160 200 240

population in thousands

high **birth rates** and falling **death rates**, characteristic of the second stage of the **demographic transition** and displayed by many **Third World** countries today. *Narrow-based* and tall pyramids indicate low birth rates and low death rates with high longevity, as typified by the developed nations in the final stage of the demographic transition. See diagram on pp. 154–5.

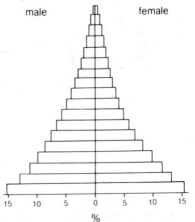

Example of a broad-based population pyramid.

male   female

15   10   5   0   5   10   15

%

Postindustrial A phrase applied to those nations which have passed through the period of heavy **industrialization**, and in which the economy is increasingly dominated by microelectronics, automation, and the service sector.

Pothole (or **sinkhole** or **swallow hole**) A feature of

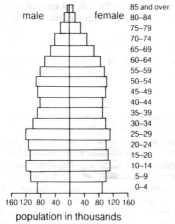

Example of a narrow-based population pyramid.

male   female

85 and over
80–84
75–79
70–74
65–69
60–64
55–59
50–54
45–49
40–44
35–39
30–34
25–29
20–24
15–20
10–14
5–9
0–4

160 120 80 40 0 40 80 120 160

population in thousands

limestone country: the solution effect of rainwater causing certain **joints** to be enlarged and thereby producing a hole of varying proportions in the limestone surface.

Surface streams may disappear down potholes to become underground watercourses. Gaping Gill in the Ingleton district of Yorkshire is a classic example of a pothole. The term pothole is also used to describe the hollows scoured in a river bed by the swirling of pebbles and small boulders in eddies.

Precipitation   Water deposited on the earth's surface in the form of, e.g., rain, snow, sleet, hail and dew.

Pre-industrial   A term used to describe the early stages of the development process. Largely agricultural economies in which the foundations for development are being established such as agricultural extension projects and improved **communications** would be described as pre-industrial. The implication of such terminology is that **industrialization** can be equated with development: while this has been true in recent history, it may be that alternative models will emerge as **resource** shortage and **pollution** become recognized globally.

Prevailing wind   The dominant wind direction of a region. In Northwest Europe for example, south-westerly winds predominate, i.e. occur more frequently than any other direction. On a global **scale**, prevailing winds occur as on the diagram on the opposite page.

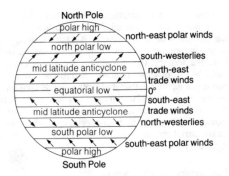

North Pole
polar high
north polar low
mid latitude anticyclone
equatorial low — 0°
mid latitude anticyclone
south polar low
polar high
South Pole

north-east polar winds
south-westerlies
north-east trade winds
south-east trade winds
north-westerlies
south-east polar winds

Primary sector That sector of the national economy which deals with the production of primary materials: **agriculture**, mining, forestry and fishing. Primary products such as these have had no processing or manufacturing involvement. The total economy comprises the primary, secondary, tertiary and quaternary sectors. The **secondary sector** includes all manufacturing and processing industries, the **tertiary sector** is dominated by **services** such as transport, retailing and the professions, and the **quaternary sector** provides information and expertise. The distribution of employment between these three sectors is a measure of the state of development of the nation: the early stages of the development process are marked by a concentration of the labour force in the primary sector, especially in agriculture, whilst the progress of development is characterized by increasing employment in firstly the secondary, and later the tertiary and quaternary sectors. It is also possible to observe regional variations within a country, e.g., the Southeast of Britain would contain a greater proportion of tertiary and quaternary workers than certain northern areas of Britain. A mining area would contain a large number of primary workers, and an industrial region would contain a large number of secondary workers.

Pull factors See **Migration**.

Pumped storage In the hydro-electricity generating industry, surplus 'off peak' electricity is used to pump water back up into the storage lake above the station, adding to the head of water for subsequent power generation.

Push factors See **Migration**.

Pyramidal peak A pointed mountain summit resulting from the headward extension of **corries** and **arêtes**. Under glacial conditions a given summit may develop corries on all sides, especially those facing north and east. As these erode into the summit, a formerly rounded profile may be changed into a pointed, steep-sided peak. The Matterhorn in the Alps is a classic example.

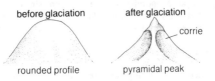

before glaciation

rounded profile

after glaciation

corrie

pyramidal peak

Pyroclasts Rocky debris emitted during a volcanic eruption, usually following a previous emission of gases and prior to the outpouring of **lava** — although many eruptions do not reach the final lava stage. Pyroclasts may comprise lumps of solidified lava from previous eruptions, as found in a volcanic **plug** or chunks of country rock, i.e. the crustal material close to the volcanic vent, or finer debris such as ash and dust. The largest pyroclasts are 'volcanic bombs', huge pieces of debris weighing up to several tons. Pebble-sized debris is referred to as *lapilli*.

Quartz One of the commonest minerals found in the earth's **crust**, and a form of silica (silicon + oxygen). Most **sandstones** are predominantly composed of quartz.

Quartzite A very hard and resistant **rock**, formed by the metamorphism of **sandstone**.

Quaternary sector That sector of the economy providing information and expertise. This includes the microchip and microelectronics industries. Highly developed economies are seeing an increasing number of their workforce employed in this sector. Compare **Primary sector, Secondary sector, Tertiary sector**.

Range The maximum distance a consumer is prepared to travel in order to purchase a good or service in a central place (see **central place theory**). The range of low order, everyday goods such as bread, newspapers and daily groceries is very short. Consequently, journeys for these goods are frequent and other people will only travel short distances. On the other hand, the range of high order specialized goods is much greater and people will therefore travel longer distances to purchase these items and their journeys will be less frequent.

Rank size rule A theory of the numerical distribution of **settlements**, such that for a given settlement system:

$$Pr = \frac{Pl}{r}$$

where:

Pr is the population of the 'r'th rank settlement.
Pl is the population of the largest settlement.

Thus, for example if Pl = 1000 then the population

of the second rank settlement will be

$$P_2 = \frac{1000}{2} = 500$$

and the population of the third rank settlement will be

$$P_3 = \frac{1000}{3} = 333$$

etc.

If a rank size prediction is made for all settlements in a system, then the graph of population against rank appears as a J-shaped curve thus:

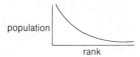

Transformed logarithmically, the graph appears thus:

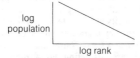

The rank size rule suggests that in any settlement system there will be very few large places and very small places — few cities and many villages. This is supportive of **central place theory**, and the fact that numerical distributions approximating to the rank size graph have been observed in reality, especially for nations mid-way through the development process such as those in central and eastern Europe, suggests that the processes implicit in Christaller's model are indeed at work. However, there are two observed deviations from the theoretical rank size distribution:

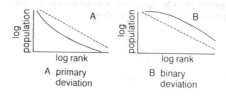

A primary deviation

B binary deviation

Primary deviation technically means that the size of the second largest city is less than half the size of the largest. Such a situation occurs when a *primate city* emerges; a rapidly growing metropolis, much larger than any other city in the system, resulting from large-scale rural to urban **migration**. Such cities as São Paulo in Brazil, Mexico City, Nairobi, Lagos are primate cities which are attracting large numbers of in-migrants. This is indicative of the early to middle stages of the development process, when development and its associated opportunites are concentrated in a national **core**. Binary deviation reflects the later stages of the development process as regional and local centres emerge: technically binary deviation means that the second largest city is more than half the size of the largest.

In a highly developed nation there may be several industrial cities or **conurbations**, large and of similar size and this will be reflected in a binary curve on the rank-size graph.

Rapid An area of broken, turbulent water in a river channel, caused by the outcropping of resistant **rocks**. If such rocks dip downstream, then rapids result; if upstream, a **waterfall** will be created.

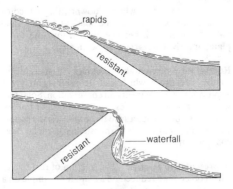

Raw materials The **resources** supplied to industries for subsequent manufacturing processes: e.g. agricultural products, minerals and timber are **raw materials**. Many **primary products** are used as raw materials.

Regeneration Renewed growth of, for example, forest after felling. Forest regeneration is crucial to the long-term stability of many **resource** systems, from shifting cultivation to commercial forestry.

Rejuvenation A fall in sea level, or a rise in the level of the land relative to the sea, will cause renewed vertical **corrasion** by rivers in their middle and lower courses.

The point at which downcutting recommences (*knickpoint*) may be marked by a **waterfall**. **Meanders** will become *incised* into the **flood plain**, and **river terraces** may be created.

Renewable/nonrenewable resources Renewable resources are those which can be used again given appropriate management and conservation. Solar and wind energy are examples, as are water and timber. Nonrenewable resources are those with a fixed supply which will eventually be exhausted: e.g. metal ores, **coal** and oil.

Representative fraction The fraction of real size to which objects are reduced on a map: for example on a

1:50,000 map, any object is shown at 1/50,000 of its real size.

Resource Any aspect of the human and physical **environments** which people find useful in the satisfying of their needs.

Revolution The passage of the earth around the sun; one revolution is completed in 365.25 days. Due to the tilt of the earth's axis (23½° from the vertical), revolution results in the sequence of seasons experienced on the earth's surface:

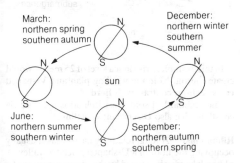

March:
northern spring
southern autumn

December:
northern winter
southern summer

sun

June:
northern summer
southern winter

September:
northern autumn
southern spring

Ria A submerged river valley, caused by a rise in sea level or a subsidence of the land relative to the sea. For example, a postglacial rise in sea level may lead to coastal submergence.

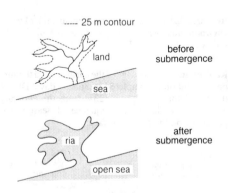

----- 25 m contour

land

sea

before submergence

ria

open sea

after submergence

In this example, a rise in sea level of 25 metres would create the ria. Note the winding plan and V-shaped **cross section**; contrast with **fjord**.

The seaboard of southwest Ireland is a classic ria coastline. See also **Discordant coastline**.

Ribbon lake A long, relatively narrow lake, usually occupying the floor of a U-shaped glaciated valley. A ribbon lake may be caused by the *over-deepening* of a section of the valley floor by glacial **abrasion**. Such a situation would occur, for example, where softer or weakened **rocks** outcrop on the valley floor. Alternatively a ribbon lake may be dammed back by a terminal or recessional **moraine**. Many of the lakes of the English Lake District — for example Windermere, Coniston Water, Wastwater — are ribbon lakes.

Rift valley A section of the earth's **crust** which has been downfaulted. The **faults** bordering the rift valley are approximately parallel. There are two main theories related to the origin of rift valleys. The first states that tensional forces within the earth's **crust** have caused a block of land to sink between parallel faults.

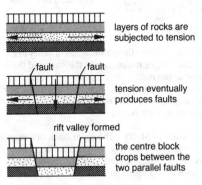

layers of rocks are subjected to tension

fault   fault

tension eventually produces faults

rift valley formed

the centre block drops between the two parallel faults

The second theory states that compression within the earth's crust has caused faulting in which two side blocks have risen up towards each other over a central block.

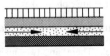

layers of rock are subjected to compression

faults develop and the outer blocks are forced over the central block

overhang will eventually be eroded

The river Rhine flows through a rift valley in Europe with the Vosges mountains to the west and the Black Forest to the east.

The most complex rift valley system in the world is that ranging from Syria in the Middle East to the river Zambesi in East Africa.

River cliff (or **bluff**) The outer bank of a **meander**. The cliff is kept steep by undercutting since river **erosion** is concentrated on the outer bank. See diagram on opposite page.

168

River's course The river's progress from its mountain origins to the sea can be divided into three major sections: the upper course, the middle course and the lower course. The characteristics of these stages are as follows:

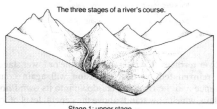

The three stages of a river's course.

Stage 1: upper stage

swift current	steep gradient
straight course	coarse load
vertical erosion	waterfalls
deep, steep sided	interlocking spurs
V-shaped valley	

Stage 2: middle stage

predominantly lateral erosion	wider valley
less steep gradient	deposition of alluvium on
reduced velocity	valley floor
meandering river	

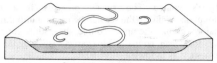

Stage 3: lower stage

some lateral erosion	deltas at mouth
deposition predominant	gradient slight
thick deposits of alluvium	velocity slow
wide valley	levées
ox bow lakes	extensive meanders

River terrace   If a river in the middle or lower stages is **rejuvenated**, vertical **corrasion** will again commence and the river will cut down into its own **flood plain**, which now stands above the general level of the river as a river terrace:

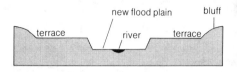

Roche moutonnée   An outcrop of resistant **rock**, sculpted as in the following diagram by the passage of a **glacier**:

170

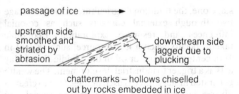

passage of ice ⟶

upstream side
smoothed and
striated by
abrasion

downstream side
jagged due to
plucking

chattermarks – hollows chiselled
out by rocks embedded in ice

Rock The solid material of the earth's **crust**. See **Igneous, Sedimentary, Metamorphic rocks**.

Rostow's model A model devised by economist W. Rostow in 1955 to describe the stages of the development process:

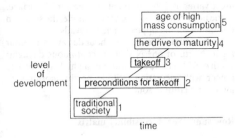

level
of
development

age of high
mass consumption ⌉5

the drive to maturity ⌉4

takeoff ⌉3

preconditions for takeoff ⌉2

traditional
society ⌉1

time

Rostow's model suggests that nations progress from

stage one, the traditional agricultural society, to stage two through external contacts such as colonial influences and **resource** exploration. Economic takeoff occurs when a key resource, such as **coal** in 19th century Britain, or oil in the Middle East today acts as a trigger to industrial development. The wealth generated in stage three is used in stage four to create a diverse industrial base and to develop social and welfare services, while stage five marks the ultimate development of the urban-industrial system, as typified by such nations as the USA.

Rotation The movement of the earth around its own axis. One rotation is completed in twenty-four hours. Due to the tilt of the earth's axis, the length of day and night varies at different points on the earth's surface. In the northern midsummer, for example, the situation illustrated on the opposite page prevails.

At the equator there is a 12-hour day and a 12-hour night. North of $66\frac{1}{2}$°N there is continuous daylight; south of $66\frac{1}{2}$°S there is continuous night. Days become longer with increasing latitude north; shorter with increasing latitude south.

Row sum See **Accessibility matrix**.

Rural depopulation The loss of population from the

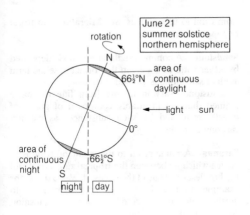

June 21
summer solstice
northern hemisphere

rotation

N

66½°N — area of continuous daylight

light   sun

0°

area of continuous night

66½°S

S

night | day

countryside as people move away from rural areas towards cities and **conurbations**. Most industrial nations have experienced a shift of population from countryside to city as the economy has evolved from agricultural origins into a predominantly urban-industrial system. Rural communities lose their most dynamic members; generally the migrating population is dominated by the younger, most active, most progressive individuals.

Rural–urban migration   The movement of people

from rural to urban areas. See **Migration** and **Rural depopulation**.

Sandstone A common **sedimentary rock**, deposited by either wind or water; in the former case the term **aeolian** sandstone is applied.

Sandstones vary in texture from fine to coarse grained, but are invariably composed of grains of **quartz**, cemented by such substances as calcium carbonate or silica.

Savanna A name given to the grassland regions of Africa which lie between the **tropical rain forest** and the hot deserts. Thus in the states of West Africa, for example, a range of vegetation exists from south to north: tropical rainforest to hot desert via a transition zone of savanna grassland:

Near the forest margin the savanna region

174

comprises extensive woodland with occasional expanses of grassland; near the desert margin the vegetation is sparse thorny scrub. Commercial **pastoral farming** is the dominant land use of the central savannas, while nomadic pastoralism prevails towards the desert margin. In South America the *Llanos* and *Campos* regions are representative of the savanna type.

Scale The size ratio represented by a map: for example on a map of scale 1:25,000 the real landscape is portrayed at 1/25,000 of its actual size.

Scarp slope The steeper of the two slopes which comprise an **escarpment** of inclined strata:

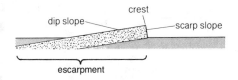

Scatter diagram A graph for displaying the relationship between two variables, as measured at a given number of observation points. For example the relationship between the population size of

settlements and the number of functions they offer might appear thus:

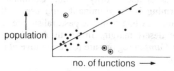

A 'line of best fit' can be plotted on the scatter to show the average trend of the relationship

⊙ Residuals – settlements which do not conform to the general relationship

Scree (or **talus**)   The accumulated **weathering** debris below a **crag** or other exposed rock face. Larger boulders will accumulate at the base of the scree, carried there by greater momentum.

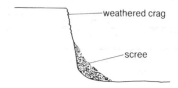

Sea breeze   See **Land breeze**.

Seafloor spreading   See **Plate tectonics**.

Secondary sector The sector of the economy which comprises manufacturing and processing industries, to be contrasted with the **primary sector** which produces **raw materials**, the **tertiary sector** which provides **services**, and the **quaternary sector** which provides information. An example of the secondary sector would be the manufacture of iron and steel.

Sector model (or **Hoyt model**) A model of urban structure developed by H. Hoyt in 1939 and based on an analysis of the land use patterns of 142 American cities.

1. central business district
2. industrial zone
3. low quality housing
4. medium quality housing
5. high quality housing

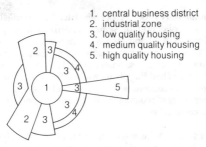

The model differs from Burgess' *concentric model* in allowing sectorial development along major lines of communication. Thus zone 2 (industry) and its accordant workers housing zone (3) will develop along, for example, a river valley which favoured canal and railway expansion. Zone 5, high quality housing, will extend along ridges of high ground or other pleasant environmental corridors. Whilst Hoyt's model is a step closer to reality, the **multiple nuclei model** is more so. No model will be an entirely successful simulation of real urban land use patterns, as the local factors responsible for the structure of a given city will be unique to that location.

Sediment The material resulting from the **weathering** and **erosion** of the landscape, and which may be reconsolidated to form **sedimentary rocks**.

Sedimentary rock A rock which has been formed by the consolidation of **sediment** derived from pre-existing rocks. **Sandstone** is a common example of a rock formed in this way; mudstone and shale are other examples. Such sedimentary rocks often show evidence of **bedding planes** which differentiate annual **deposition** sequences. **Limestone** and **evaporites** are other types of sedimentary rock derived from organic and chemical origins.

Self dune A linear sand dune, the ridge of sand lying parallel to the prevailing wind direction. Contrast with the crescent-shaped **Barchan** dune.

Serac A pinnacle of ice, formed by the tumbling and shearing of a **glacier** at an **ice fall**, i.e. the broken ice associated with a change in **gradient** of the valley floor.

Services Any urban **functions** such as banking, insurance, transport and distribution. Hence consumers may purchase or make use of 'goods and services' in urban centres, the number and range of such goods and services available being in proportion to the size of the urban place.

Settlement Any location chosen by people as a permanent or semipermanent dwelling place. Hence settlements may vary from an individual farmhouse in an agricultural landscape to a **conurbation** of several millions of people in an urban/industrial region.

Settlement hierarchy A series of size orders in a **settlement** system, the number of settlements in each order descending as the hierarchy is ascended.

Order	Number
4th order:	1
3rd order:	3
2nd order:	9
1st order:	27

Generally, a settlement hierarchy will comprise many low order places (or villages), and very few high order places (or cities) with a number of intermediate orders between the two extremes. The logic behind the hierarchy concept is formalized in Walter Christaller's **central place theory**. Christaller suggested 'k' values to indicate the numerical relationship between one order and another; for example in his k = 3 hierarchy the number of settlements is three times fewer at each successive higher order.

Such a hierarchy is, of course, entirely theoretical but empirical studies do suggest some evidence of hierarchical organization.

Shading map (or chorpleth map) A map of which shading of varying intensity is used. For example, the pattern of **population densities** in a region can be shown by means of a shading system.

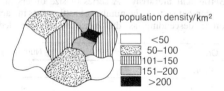

population density/km²

☐ <50
▒ 50–100
▥ 101–150
▨ 151–200
■ >200

Shaft mine A vertical mine as opposed to drift or adit mines.

Shanty town An area of unplanned, random urban development often around the edge of a city and generally occurring in **Third World** nations: cities such as São Paulo, Mexico City, Nairobi, Calcutta, Lagos are examples of this. The shanty town is characterized by high density/low quality dwellings often constructed from the simplest materials such as scrap wood, corrugated iron and plastic sheeting — and by the lack of standard services such as sewerage and water supply, power supplies and refuse collection. The shanty town is the make-shift home of the relatively recent rural-urban migrant: unemployment tends to be high since the supply of urban employment is considerably less than the demand. However, the conventional image of the shanty town is misleading: the fact that such **settlements** survive is a measure of success; many occupants gradually improve their property as they become established in the urban sector, and the **informal economy** thrives.

The shanty town is a major element of the structure of many Third World cities. There are major contrasts between the standard models of the structure of Western cities and the typical land use pattern of a Third World city, as summarized thus:

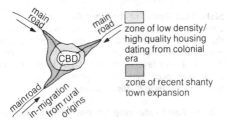

zone of low density/high quality housing dating from colonial era

zone of recent shanty town expansion

Shifting cultivation   See **Bush fallowing**.

Shoreface terrace   A bank of **sediment** accumulating at the change of slope which marks the limit of marine **wave-cut platform**.

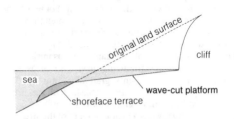

Material removed from the retreating cliff base is transported by the **undertow** off the wave-cut platform to be deposited in deeper water offshore.

Sial   That section of the lithosphere or **crust** which comprises the relatively lighter, acidic material such as those **rocks** rich in silica and aluminium: the term usually refers to continental material which lies on a deeper, denser layer — the **sima**. The terms sial and sima have been generally superseded by the terminology of **plate tectonics** theory.

Silage   Any **fodder crop** harvested whilst still green.

The crop is kept succulent by partial fementation in a *silo*.

Sill **1.** An igneous intrusion of roughly horizontal disposition. See **Igneous rocks**.
2. The lip of a **corrie**. See diagram.

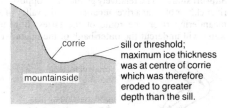

corrie

sill or threshold;
maximum ice thickness
was at centre of corrie
which was therefore
eroded to greater
depth than the sill.

mountainside

Silt Fine **sediment**, the component particles of which have a mean diameter of between 0.002 mm and 0.02 mm.

Sima That section of lithosphere or **crust** which forms the ocean floor and which underlies the continents. It is mainly of basic composition and contains heavier material than the **sial**.

Sinkhole See **Pothole**.

Slate Metamorphosed shale or **clay**. Slate is a dense,

183

fire-grained **rock** distinguished by the characteristic of *perfect cleavage*; i.e. it can be split along a perfectly smooth plane.

Slip   The amount of vertical displacement of **strata** at a **fault**.

Slip-off slope   The relatively gentle slope opposite a river cliff or **bluff** at a river **meander**. In the valley of a meandering river, the route of the fastest flowing water will undercut the outer bank of the **meander**.

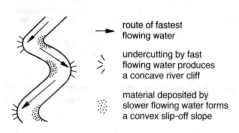

route of fastest flowing water

undercutting by fast flowing water produces a concave river cliff

material deposited by slower flowing water forms a convex slip-off slope

Eroded material from this bank may be deposited on the inner bank of the meander where the water is flowing at a slower speed. This deposited material forms a convex bank called a 'slip-off slope'. Continued deposition will eventually cause the meander to be cut off completely from the main channel and an **oxbow lake** will be formed.

Smog A mixture of smoke and fog associated with industrial areas, creating an unhealthy **atmosphere**. In 1952 a four-day smog in London left 4000 people dead or dying. Since then, many cities have had to introduce *clean-air zones*.

Snout The end of a **glacier**; strictly the point at which wasting by melting exceeds the rate of supply of ice from up-valley. A glacier snout is characterized by a dissected, discoloured appearance caused by the action of meltwater and the washing out of **moraine**.

Snow line The altitude above which permanent snow exists, and below which any snow that falls will not persist during the summer months. The altitude of the snow line varies with **latitude**: at the equator it is of the order of 5000 metres, in the Alps it is approximately 3000 metres, and at the Poles the snow line is at sea level.

Soil The loose material which forms the upper-most layer of the earth's surface, composed of the **inorganic fraction**, i.e. material derived from the **weathering** of bedrock, and the **organic fraction** — that is material derived from the decay of vegetable matter. The ultimate form of the organic fraction is humus. Soil also contains minerals and trace elements and air and water held within the soil structure. There are three broad categories of soil: *zonal*, i.e. the result of a specific combination of climatic and vegetative conditions; *intrazonal*, i.e. the result of local peculiarities of **environment** such as water-logging or unusual parent material; and *azonal*, i.e. little-developed soils which occur on such surfaces as **scree, alluvium** or sand dunes.

Soil erosion The accelerated breakdown and removal of **soil** due to poor management. Soil erosion is particularly a problem in harsh **environments**. For example, in areas of steep slope, heavy seasonal rainfall, or strong dry-season winds, the soil may be rapidly eroded once protective vegetation has been removed.

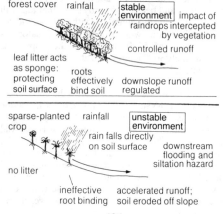

forest cover   rainfall   stable environment   impact of raindrops intercepted by vegetation

controlled runoff

leaf litter acts as sponge: protecting soil surface   roots effectively bind soil   downslope runoff regulated

sparse-planted crop   rainfall   unstable environment

rain falls directly on soil surface   downstream flooding and siltation hazard

no litter

ineffective root binding   accelerated runoff; soil eroded off slope

Soil erosion by rainwater can be classified as follows: *splash erosion*, whereby the soil is pulverized by the impact of heavy raindrops and hailstones, as in a convectional storm; *sheet erosion*, whereby in a heavy storm a surface film of water develops, which will flow downslope carrying surface soil as it moves; and *rill/gully erosion*, whereby any surface depression concentrates runoff which quickly develops into channel flow, cutting a steep-sided valley as it runs off. Soil erosion by wind occurs on extensive flatlands which are subject to a windy, dry season for part of the year. The upper soil surface becomes loose and susceptible to wind erosion due to lack of moisture.

Soil profile The sequence of layers or **horizons** usually seen in an exposed soil section.

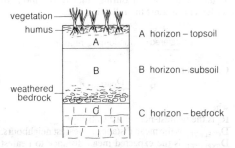

Solar power Heat radiation from the sun converted

187

into electricity. Solar power is an example of a renewable source of energy (see **Renewable/nonrenewable resources**).

Solifluction See **Periglacial features**.

Space The geographer's term for area; the context within which distributions and patterns occur.

Spatial analysis The description and explanation of distributions of people and their activities in space.

Spatial distribution The pattern of locations of, for example, population or **settlement** in a region. Geographers employ various methods for describing distributions, for example *nearest neighbour analysis*, a technique for measuring the distribution of settlements against the following scale:

Calculation:

$$R_n = \frac{D_{OBSERVED}}{D_{EXPECTED}}$$

where:

R_n is the nearest-neighbour statistic.
$D_{OBSERVED}$ is the mean distance to nearest neighbours.
$D_{EXPECTED}$ is the expected mean distance to nearest neighbours if the distribution were random.

$$D_{EXP} = \frac{1}{2\sqrt{A}}$$

where A is the density of settlements per unit area.

Spearman rank correlation coefficient   See **Correlation**.

Sphere of influence   The area, surrounding a **settlement**, from which consumers will travel in order to obtain goods and **services**. The size of the sphere of influence will vary according to the size of the settlement and the number and type of functions available there. For example, a neighbourhood shopping centre will have a *small* sphere of influence, i.e., people will not travel great distances to the centre as the range of goods and services will be limited. The central business district (**CBD**) of a town or city, however, will have a *large* sphere of influence. This will mean that people will travel from greater distances to the CBD, as the range of goods and services is much wider. See **Gravity model**.

Spit   A low, narrow bank of sand and shingle built out into an **estuary** by the process of **longshore drift**. The diagram over page of Spurn Head, Humberside is a classic example.

The **sediment** carried down the coast by longshore drift is deposited at the break in the coastline caused

by the Humber estuary. Relatively shallow water, and the slowing of longshore drift by the counter-current of the Humber have led to the **deposition** of the marine **load**.

The end of Spurn Head is 'hooked' by the action of waves swinging into the Humber estuary from the open North Sea.

Spring A resurgence of an underground stream at the surface, often occurring where **impermeable rocks** underlie **permeable strata**:

Spring line   A series of **springs** emerging along the food of an **escarpment** of **permeable rock**:

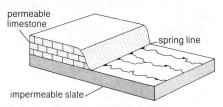

Squatter settlement   An area of peripheral urban settlement in which the residents occupy land to which they have no legal title. See **Shanty town**.

Stack   A coastal feature resulting from the collapse of a **natural arch**. The stack represents the more resistant **strata**, while softer material has been worn away by **weathering** and marine **erosion**.

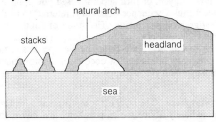

Stalactite   A column of calcium carbonate hanging from the roof of a **limestone** cavern: as water passes through the limestone it dissolves a certain proportion, which is then precipitated by evaporation of water droplets dripping from the cavern roof. These drops splashing on the floor of the cavern further evaporate to precipitate further calcium carbonate as a **stalagmite**.

Stalagmite   A column growing from the cavern floor. **Stalactites** and stalagmites may meet, forming a column.

Staple diet   The basic foodstuff which comprises the daily meals of a given people; in South East Asia rice is the staple food; in many parts of Africa the staple food is maize.

Strata   Layers of **rock** superimposed one upon the other; thus a sequence of strata is referred to as the *stratigraphy* of a region. The study of stratigraphy enables scientists to reconstruct the geological history of a region; this may be a complex process if the stratigraphy has been distrubed by earth movements such as folding or faulting.

Stratosphere   The layer of the **atmosphere** which lies immediately above the troposphere, and below the mesosphere and ionosphere. Within the stratosphere, temperatures increase with altitude. The boundary between the stratosphere and the troposphere is known as the *tropopause*.

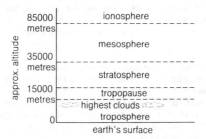

Stratus Layer-cloud of uniform grey appearance, often associated with the warm sector of a **depression**. Stratus is a type of low cloud and may hang as mist over mountain tops; broken stratus is referred to as *fractostratus*.

Striations The grooves and scratches left on bare **rock** surfaces by the passage of a **glacier**. Debris embedded in the glacier scores the surface over which it passes. Striations may thus be a guide to the direction of ice movement.

Strip cultivation A method of **soil** conservation whereby different crops are planted as a series of strips, often following **contours** around a hillside. The purpose of such a sequence of cultivation is to arrest downslope movement of soil, especially if one of the crops tends to expose the soil surface (e.g. maize). Alternate strips may be planted with grass which effectively binds the soil and acts as a brake on run off.

Subcrustal convection currents Patterns of flow in the **mantle** which are responsible for the movement of plates of the earth's **crust** (see **plate tectonics**). Such currents are produced by the heat source of the earth's core; as they ascend and spread out beneath the plates of the crust they drag the plates in the direction of current movement.

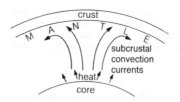

Subsoil See **Soil profile**.

Subsidiary cone A volcanic cone which develops

within the **caldera** of a previous larger cone which was blown away during an eruption.

Subsistence agriculture A system of **agriculture** in which the farmers produce exclusively for their own consumption, as contrasted with **commercial agriculture** where the farmer produces purely for sale at the market.

Surface runoff That proportion of rainfall received at the earth's surface which runs off either as channel flow or overland flow, as distinct from that proportion which percolates into the **soil**, or evaporates back into the **atmosphere**.

Swallow hole See **Pothole**.

Swash The rush of water up the beach as a wave breaks. See also **Backwash** and **Longshore drift**.

Syncline A trough in folded **strata**; the opposite of **anticline**:

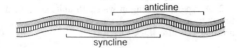

Taiga The extensive coniferous forests of Siberia and Canada, lying immediately south of the arctic

tundra. Within the taiga there are many lakes, marshes and swamps, the latter often resulting from springtime thaw occurring over permafrost and while river courses to the north are still frozen.

Tarn The postglacial lake which often occupies a **corrie**.

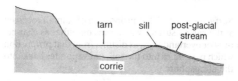

Temperate climate Climates of mid-latitudes, intermediate between the cold polar and hot tropical extremes are referred to as temperate.

Terminal moraine See **Moraine**.

Terrace 1. See **River terrace**.
2. A type of housing in which all units are joined together, as opposed to detached or semi-detached housing. See also **Back-to-back**.

Terracing A means of **soil** conservation and land utilization whereby steep hillsides are engineered into a series of flat ledges which can be used for **agriculture**.

Terracing reaches its most sophisticated and intricate form in the fertile volcanic hills of Java, Indonesia, where a subsistence rice economy prevails.

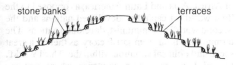

Tertiary sector That sector of the economy which provides **services** such as transport, finance and retailing as opposed to the **primary sector** which provides **raw materials**, the **secondary sector** which processes and manufactures products, and the **quaternary sector** which provides information and expertise. In highly developed economies the tertiary sector is the dominant employer, though in recent years numbers employed in the quaternary sector have gone up due to increased use of the microchip.

Textile industry The manufacture of cloth and related materials; traditionally the textile industry used wool, cotton, flax and other natural products as its **raw materials**, but this sector has declined with the rise in use of **artificial fibres** derived from oil and other hydrocarbons. Nylon, rayon and Terylene are examples of such fibres.

Thermal power station An electricity-generating

plant which burns **coal**, oil or natural gas to produce steam to drive turbines.

Third World A collective term for the poor nations of Africa, Asia and Latin America, as opposed to the 'first world' of capitalist, developed nations and the 'second world' of communist, developed nations. The terminology is far from satisfactory, as there is a great social and political variation within the 'Third World'. Indeed there are some countries where extreme poverty prevails which could be regarded as a 'fourth' group.

Threshold See **Sill (2)**.

Threshold population The minimum population required in a **sphere of influence** to sustain a particular good or service offered in a **central place**. The size of the threshold population increases with progressively more specialized goods and **services** for which individual demand is less frequent.

Tidal limit The point upstream of which there is no tidal rise and fall in river level.

Tidal range The mean difference in water level between high and low tides at a given location.

Tombolo A **spit** which extends to join an island to

the mainland, as in the case of Chesil Beach, Portland Island, southern England:

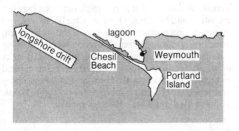

Topography The composition of the visible landscape, comprising both physical (e.g. relief, **drainage**, vegetation) and human (e.g. roads, railways, **settlements**) features.

Topsoil The uppermost layer of **soil**, more rich in organic matter than the underlying **subsoil**. See **Soil profile**.

Transhumance The practice whereby herds of farm animals are moved between regions of different climates. Pastoral farmers (see **pastoral farming**) take their herds from valley pastures in the winter to mountain pastures in the summer. Frequently farmers will live in mountain huts during the summer in

199

mountainous regions such as the Himalayas and the Alps.

Transition zone   That part of the inner city where old industrial and residential land uses are being replaced by newer functions in the process of redevelopment. Specifically, old industrial revolution **terrace** housing is being demolished to make way for new high density housing; derelict industrial property is being cleared to create new land for **CBD** expansion, landscaping and amenity. However, much derelict land remains. See also **Twilight Zone, Burgess model**.

Transpiration   The process whereby plants give off water vapour via the stomata of their leaves. Water taken up by roots is thus returned to the **atmosphere**.

Tropical rainforest   The dense forest cover of the equatorial regions, reaching its greatest extent in the Amazon Basin of South America, the Zaire Basin of Africa, and in parts of South East Asia and Indonesia. The lush forest is a response to optimum climatic conditions (high temperatures and abundant moisture); not to soil fertility. Tropical soils are, in fact, generally poor. Newly germinated seedlings grow rapidly upwards in search of light in the dense forest cover, before branching out having reached the forest canopy. Most trees are very shallow rooted, and are often buttressed to provide support. The richness and diversity of the forest itself gives rise to a similarly varied fauna.

The typical characteristics of the tropical rainforest can be summarized thus:

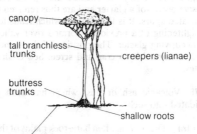

canopy

tall branchless trunks

creepers (lianae)

buttress trunks

shallow roots

intense bacterial activity breaks down fallen leaves etc. to return nutrients to soil surface for immediate uptake by roots.
Soils themselves are infertile: the nutrient cycle is concentrated in the vegetation and top few inches of soil.

Trough An area of low pressure, not sufficiently well-defined to be regarded as a **depression**.

Trough end wall The steep rear wall of a **U-shaped valley**, formed where coalescing **corrie glaciers** cause an increase in **erosion** and consequent deepening of the glacial valley.

Truncated spur   Formerly **interlocking spurs** in a V-shaped river valley are sheared off by the greater erosive power of a **glacier** and are thus referred to as truncated spurs. It is this process which leads to the straightening of a previously formed river valley by an occupying glacier. The position of the truncated spur is marked by a **crag** and **scree**. See **U-shaped valley**.

Tuff   Volcanic ash or dust which has been consolidated into **rock**.

Tundra   The barren, often bare-rock plains of the far north of North America and Eurasia where subarctic conditions prevail and where, as a result, vegetation is restricted to low-growing, hardy shrubs and mosses and lichens. *Permafrost* conditions result in poor **drainage** which leads to marsh and swamp during the short summer.

Twilight zone   Another term for the **transition zone** in urban structure, implying especially the run-down, semiderelict nature of much of this area.

Undernutrition   A lack of sufficient quantity of food, as distinct from **malnutrition** which is a consequence of unbalanced diet. Many of the world's poorest people suffer both undernutrition and malnutrition. As the world's population rises at a faster rate than that of food supply, problems of undernutrition are becoming worse in many parts of Latin America, Africa and Asia.

Undertow The counter-current to water breaking onshore as waves. The undertow is responsible for the removal of eroded material from the **wave-cut platform**, to be deposited as the **shoreface terrace**. The undertow operates at a much larger scale than **backwash**.

Urban decay The process of deterioration in the **infrastructure** of parts of the city — especially in the old industrial cities of, for example, the English North and Midlands.

Parts of the inner city are especially decayed: old Victorian **terrace** housing, mills and other traditional industrial installations, disused canals and warehouses. Much of the decay results from neglect as the focus of the urban system moves away from these areas — for example towards new peripheral locations for industry, newer housing in the outer suburbs and new **communications** bypassing the city. Many cities have undertaken 'face-lift' schemes to improve the decayed **environments**, either by demolition and landscaping, or by renovating old property for new uses.

Urbanization The process by which a national population becomes predominantly urban, through a **migration** of people from the countryside to cities and a shift from agricultural to industrial employment. Urbanization is thus an important element of the development process. See also **Migration** and **Rank size rule.**

Urban sprawl The growth in extent of an urban area in response to improvements in transport and rising incomes, both of which allow a greater physical separation of home and work. The sprawling outer suburbs of cities today are a result of almost-universal car ownership; in previous years similar, though smaller, expansions of the urban area have resulted from the development of the urban railway, the tram and the bus. The sequence can be summarized thus (UK cities):

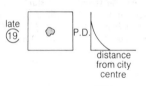

date	urban area	urban population density gradient	conditions
late 19		P.D. \\ distance from city centre	low mobility, majority of urban population walked to work

mid
20

spreading city
in response
to developing
public
transport

1980's

high mobility
due to higher
incomes
and car
ownership

U-shaped valley A glaciated valley, characteristically straight in plan and U-shaped in **cross section**.

The winding V-shaped valley of a river's course is modified into a U-shaped valley by the greater erosive power of a **glacier**. The major features of a U-shaped valley in post-glacial times are summarized in the diagram over page.

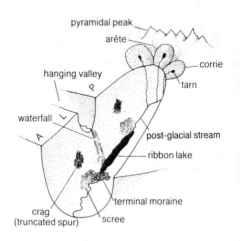

pyramidal peak

arête

corrie

tarn

hanging valley

waterfall

post-glacial stream

ribbon lake

terminal moraine

crag
(truncated spur)

scree

Viscosity A measure of the fluidity of, for example, **lava**: viscous lava is sticky, flows slowly, and congeals rapidly. Non-viscous lava is very fluid, flows quickly and congeals slowly.

Vicious cycle of poverty The poverty trap in which much of the population of the **Third World** finds itself. Poor farmers cannot invest in their land through improved seed or fertilizer; **yields** thus remain low, there is no surplus for sale at market and so the poverty continues. In the absence of credit or grand aid, there is no way in which the farmer can break out of the cycle.

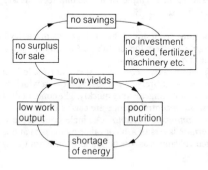

Given that poor yields also lead to a food shortage, the farmer will suffer poor nutrition and may have to borrow money even to secure enough food, let alone invest in the land. It is not only necessary to provide the financial means for the farmer to improve his or her livelihood; it is also important to provide the correct institutional context for progress — for example land reform may be necessary to ensure security of tenure.

Volcanic rock A category of **igneous rocks** which

comprises those rocks formed from **magma** which has reached the earth's surface. (This is to be contrasted with **plutonic rocks** which form below the surface.) **Basalt** is an example of a volcanic rock, as are all solidified lavas.

Volcano A fissure in the earth's **crust**, through which **magma** reaches the earth's surface. There are four main types of volcano:

(a) *Acid lava cone* A very steep-sided cone composed entirely of viscous, acidic lava which flows slowly and congeals very quickly. The cones of the Puy District in central France are classic examples.

(b) *Composite volcano* A single cone comprising alternate layers of ash (or other pyroclasts) and lava. Such volcanoes as Vesuvius and Etna are of this type.

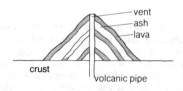

(c) *Fissure volcano* Eruptions along a linear fracture in the crust, rather than from a single cone, are called fissure eruptions and are generally of a quiet, unexplosive nature, relating to the non-viscous **lavas** which are usually produced. Many eruptions in Iceland are of this type.

(d) *Shield volcano* (or *lava cone*) A volcano composed of very basic, non-viscous lava which flows quickly and congeals slowly, producing a very gently sloping cone; e.g. Mona Loa, Hawaii:

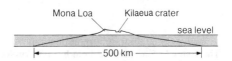

Von Thünen theory J. H. Von Thünen's early-nineteenth century model of the distribution of agricultural land use around a town, and still the basis for many investigations of such land use patterns. Von Thünen envisaged a single market town, surrounded by a region supplying agricultural produce. Physical and economic characteristics in this **hinterland** are regarded as uniform: an isotropic surface inhabited by rational 'economic man'.

Such conditions of course do not exist in reality, but the value of a model like Von Thünen's is that it sweeps away the clutter of reality and allows us to observe basic processes at work. Von Thünen postulated the following land use system, given his simplifying assumptions, illustrated in the diagram over page.

1 market gardening
2 dairying
3 arable
4 rough grazing
5 wilderness

The zones 1—5 are derived from **bid rent curves** thus:

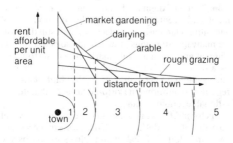

The implication of the theory is that the distribution

of land uses around the town will depend upon a variety of factors: perishability, transport costs and land area requirements.

Highly perishable produce such as salad crops must be grown close to town to ensure freshness at market and to minimize the transport costs necessitated by frequent marketing. The location of market gardens close to the town is also explained by the fact that land area requirements are small and therefore the bid rent per unit area can be high. Market gardeners can achieve high productivity from small areas by the intensive nature of their farming, through investment in greenhouses, fertilizer and other inputs. Land uses with progressively larger land area requirements must locate further away from the town; a farmer needing many units of land will only be able to bid a lower rent per unit area. Land uses at progressively greater distance from town are also characterized by a decreasing frequency of marketing of products, and by decreasing investment on the land. In reality such patterns are constrained by variations in the physical **environment** and in human behaviour; government policy also affects farming practices and the distribution of land uses.

Vulcanicity A collective term for those processes which involve the intrusion of **magma** into the **crust**, or the extrusion of such molten material onto the earth's surface.

Wadi A dry watercourse in an arid region: occasional rainstorms in the desert may cause a temporary stream to appear in a wadi.

Warm front See **Depression**.

Warm sector See **Depression**.

Waterfall An irregularity in the long profile of a river's course, usually located in the upper stage. A waterfall occurs where alternating hard and soft **strata** outcrop; the softer material is eroded more quickly and a sharp change of **gradient** is created. A waterfall may also be the result of **rejuvenation**, caused by the re-establishment of vigorous vertical corrasion.

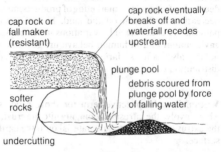

cap rock or fall maker (resistant)

cap rock eventually breaks off and waterfall recedes upstream

plunge pool

debris scoured from plunge pool by force of falling water

softer rocks

undercutting

Water gap In chalk **escarpments**, a valley which has been eroded below the depth of the **water table**, and which therefore contains a permanent stream.

Watershed The boundary, often a ridge of high ground, between two river basins:

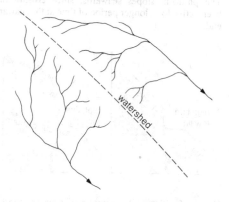

Water table The level below which the ground is permanently saturated. The water table is thus the upper surface of the **groundwater**. In areas where **permeable** rock predominates, the water table may be at some considerable depth. In periods of high rainfall the level of the water table will rise; it will fall in protracted dry spells.

Wave-cut platform (or **abrasion platform**) A gently sloping bench eroded by the sea along a coastline. As the **cliff** recedes, the abrasion platform is created. The platform slopes seawards, since **erosion** has been active for a longer period of time at the seaward end.

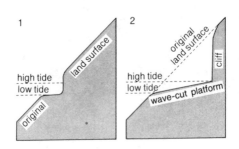

Wave refraction The bending of waves around a **headland**. The shorewards movement of water in contact with the headlands is slowed, while open water in the middle of the bay moves on unimpeded. A vertical view of wave refraction is illustrated in the diagram over page.

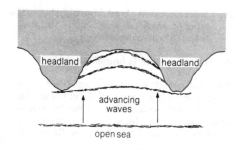

Weather The day-to-day conditions of, e.g., rainfall, temperature, pressure, as experienced at a particular location. Contrast with climate, which is a set of long-term, average atmospheric conditions. Thus the climate of, for example, a location in the Sahara desert, will be 'hot and dry all the year round', but there will be occasions when violent weather in the form of conventional rainstorms occurs.

Weathering The breakdown of rocks in situ; contrasted with **erosion** in that no large-scale transport of the denuded material is involved. Weathering processes include **exfoliation, nivation** and chemical activity such as the dissolving of **limestone** by rainwater. Biological activity, for example by tree roots and earthworms, also contributes towards the breakdown of bedrock. Thus three types of weathering are identified: mechanical (or physical), chemical and biological.

Weber's theory See **Industrial location**.

White-collar worker A worker who is not a manual worker and who does not work in 'dirty conditions'. The term 'white-collar' derives from the idea of a white shirt as worn by, e.g., clerical workers and professional people. Compare **Blue-collar worker**.

White ice Ice from which air has not been totally expelled (see **corrie**). Contrast with **blue ice**, found at greater depth in a corrie icefield, and from which air has been expelled by compression.

Wind gap A **dry valley**, for example in a chalk **escarpment**, now standing above the **water table** but formed at a time when the water table was higher or when the ground was frozen.

Yield The productivity of land as measured by the weight or volume of produce per unit area. Agricultural yields are usually expressed per hectare.

Yardang Long, roughly parallel ridges of **rock** in arid and semi-arid regions; the ridges are undercut by wind **erosion** and the corridors between them are swept clear of sand by the wind. The ridges are oriented in the direction of the prevailing wind. The term 'yardang' comes from Central Asia.

Zeugen Pedestal rocks in arid regions; wind **erosion** is concentrated near the ground where **corrasion** by wind-borne sand is most active. This leads to undercutting and the pedestal profile emerges:

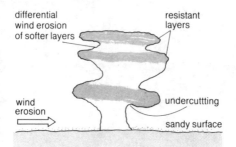

differential wind erosion of softer layers

resistant layers

wind erosion

undercuttting

sandy surface

Appendix I

Major source:
World Population Data Sheet,
Population Reference Bureau,
Washington DC.

Selected demographic data for a sample of nations

Nation	Persons per sq. km of arable land	Life expectancy (years)	Population under age 15 years (% of total)	Urban population (% of total)	Gross national product per capita ($)	Rate of natural increase in pop; % per year	Population estimate 1981 (millions)
Europe							
France	169	73	23	78	9940	0.4	53.9
Greece	105	73	23	65	3890	0.7	9.6
Hungary	160	70	21	54	3850	0.2	10.7
Italy	326	73	22	69	5240	0.2	57.2
Netherlands	694	75	23	88	10240	0.4	14.2
Poland	189	71	24	57	3830	1.0	36.0
Spain	120	73	26	74	4340	0.8	37.8
Sweden	223	75	20	83	11920	0.1	8.3
West Germany	465	72	20	92	11730	-0.2	61.3

Nation	Persons per sq. km of arable land	Life expectancy (years)	Population under age 15 years (% of total)	Urban population (% of total)	Gross national product per capita ($)	Rate of natural increase in pop; % per year	Population estimate 1981 (millions)
North America							
Canada	35	74	25	76	9650	0.8	24.1
United States	53	74	22	74	10820	0.7	229.8
Central America							
Mexico	71	65	42	67	1590	2.5	69.3
Nicaragua	52	55	48	53	660	3.4	2.5
Panama	109	70	43	51	1350	2.2	1.9

221

Nation	Persons per sq. km of arable land	Life expectancy (years)	Population under age 15 years (% of total)	Urban population (% of total)	Gross national product per capita ($)	Rate of natural increase in pop; % per year	Population estimate 1981 (millions)
South America							
Argentina	16	69	28	82	2280	1.6	28.2
Brazil	58	64	41	61	1690	2.4	121.4
Colombia	120	62	41	60	1010	2.1	27.8
Ecuador	159	60	45	43	1050	3.1	8.2
Venezuela	70	66	42	75	3130	3.0	15.5
Australia	3	73	27	86	9100	0.9	14.8
USSR	44	69	26	65	4110	0.8	268

Nation	Persons per sq. km of arable land	Life expectancy (years)	Population under age 15 years (% of total)	Urban population (% of total)	Gross national product per capita ($)	Rate of natural increase in pop; % per year	Population estimate 1981 (millions)
Africa							
Algeria	44	56	48	61	1580	3.2	19.3
Egypt	1533	55	41	45	460	3.0	43.5
Ghana	90	48	46	31	400	3.1	12.0
Kenya	273	53	50	14	380	3.9	16.5
Lesotho	59	50	41	5	340	2.4	1.4

	Mali	Nigeria	South Arica	Tanzania	Zaire	Zambia
Persons per sq. km of arable land	21	178	30	38	97	104
Life expectancy (years)	42	48	60	50	46	53
Population under age 15 years (% of total)	45	47	42	46	44	49
Urban population (% of total)	17	20	50	12	30	20
Gross national product per capita ($)	140	670	1720	270	260	510
Rate of natural increase in pop; % per year	2.8	3.2	2.4	3.0	2.8	3.2
Population estimate 1981 (millions)	6.8	79.9	29.0	19.2	30.1	6.0
Nation						

Nation	Persons per sq. km of arable land	Life expectancy (years)	Population under age 15 years (% of total)	Urban population (% of total)	Gross national product per capita ($)	Rate of natural increase in pop; % per year	Population estimate 1981 (millions)
Asia							
Afghanistan	28	42	45	15	170	2.7	16.4
China	309	68	32	25	230	1.2	985.0
India	381	52	41	22	190	2.1	688.6
Iran	66	58	44	50	?	3.0	39.8
Japan	2145	76	24	76	8800	0.8	117.8

Nation	Jordan	Malaysia	Philippines	Saudi Arabia	Thailand	Turkey
Persons per sq. km of arable land	223	220	538	12	273	83
Life expectancy (years)	56	61	61	48	61	61
Population under age 15 years (% of total)	52	41	43	45	43	39
Urban population (% of total)	42	29	36	67	14	47
Gross national product per capita ($)	1180	1320	600	7370	590	1330
Rate of natural increase in pop; % per year	3.3	2.3	2.4	3.0	2.0	2.2
Population estimate 1981 (millions)	3.3	14.3	48.9	10.4	48.6	46.2

226

Appendix II

National Capitals

Nation	*Capital*
Afghanistan	Kabul
Albania	Tirana
Algeria	Algiers
Andorra	Andorra La Vella
Angola	Luanda
Anguilla	The Valley
Antigua	St. John's
Argentina	Buenos Aires
Australia	Canberra
Austria	Vienna
Bahamas	Nassau
Bahrain	Manama
Bangladesh	Dacca
Barbados	Bridgetown
Belgium	Brussels
Belize	Belmopan
Benin	Porto Novo
Bermuda	Hamilton
Bhutan	Thimbu
Bolivia	La Paz
Botswana	Gaborone
Brazil	Brasilia
Brunei	Bandar Seri Begawan
Bulgaria	Sofia

Nation	Capital
Burma	Rangoon
Burundi	Bujumbura
Cambodia	Phnom Penh
Cameroon	Yaoundé
Canada	Ottawa
Cayman Islands	Georgetown
Central African Republic	Bangui
Chad	Ndjamena
Chile	Santiago
China	Peking
Colombia	Bogotá
Comoro Islands	Moroni
Congo	Brazzaville
Costa Rica	San José
Cuba	Havana
Cyprus	Nicosia
Czechoslovakia	Prague
Denmark	Copenhagen
Djibouti	Djibouti
Dominica	Roseau
Dominican Republic	Santo Domingo
Ecuador	Quito
Egypt	Cairo
El Salvador	San Salvador
Equatorial Guinea	Rey Malabo
Ethiopia	Addis Ababa
Falkland Islands	Stanley
Fiji	Suva

Nation	Capital
Finland	Helsinki
France	Paris
French Guiana	Cayenne
Gabon	Libreville
Gambia	Banjul
Germany (East)	East Berlin
Germany (West)	Bonn
Ghana	Accra
Greece	Athens
Greenland	Godthaab
Grenada	St. George's
Guadeloupe	Basse Terre
Guam	Agana
Guatemala	Guatemala City
Guinea	Conakry
Guinea-Bissau	Bissau
Guyana	Georgetown
Haiti	Port-au-Prince
Honduras	Tegucigalpa
Hong Kong	Victoria
Hungary	Budapest
Iceland	Reykjavik
India	New Delhi
Indonesia	Djakarta
Iran	Tehran
Iraq	Baghdad
Ireland	Dublin
Israel	Jerusalem
Italy	Rome

Nation	Capital
Ivory Coast	Abidjan
Jamaica	Kingston
Japan	Tokyo
Jordan	Amman
Kenya	Nairobi
Kiribati	Tarawa
Korea (North)	Pyongyang
Korea (South)	Seoul
Kuwait	Kuwait City
Laos	Vientiane
Lebanon	Beirut
Lesotho	Maseru
Liberia	Monrovia
Libya	Tripoli
Luxembourg	Luxembourg
Macau	Macau
Madagascar	Antananarivo
Malawi	Lilongwe
Malaysia	Kuala Lumpur
Maldive Islands	Malé
Mali	Bamako
Malta	Valletta
Martinique	Fort-de-France
Mauritania	Nouakchott
Mauritius	Port Louis
Mayotte	Dzaoudzi
Mexico	Mexico City
Mongolia	Ulan Bator
Montserrat	Plymouth

Nation	Capital
Morocco	Rabat
Mozambique	Maputo
Namibia	Windhoek
Nepal	Katmandu
Netherlands	The Hague
New Caledonia	Nouméa
New Zealand	Wellington
Nicaragua	Managua
Niger	Niamey
Nigeria	Lagos
Norway	Oslo
Oman	Muscat
Pakistan	Islamabad
Panama	Panama City
Papua New Guinea	Port Moresby
Paraguay	Asunción
Peru	Lima
Philippines	Manila
Poland	Warsaw
Portugal	Lisbon
Puerto Rico	San Juan
Qatar	Doha
Reunion	St. Denis
Romania	Bucharest
Rwanda	Kigali
Saudi Arabia	Riyadh
Senegal	Dakar
Seychelles	Victoria
Sierra Leone	Freetown

Nation	Capital
Singapore	Singapore
Solomon Islands	Honiara
Somali Republic	Mogadishu
South Africa	Pretoria
Spain	Madrid
Sri Lanka	Colombo
Sudan	Khartoum
Surinam	Paramaribo
Swaziland	Mbabane
Sweden	Stockholm
Switzerland	Berne
Syria	Damascus
Taiwan	Taipei
Tanzania	Dar es Salaam
Thailand	Bangkok
Togo	Lomé
Tonga	Nuku'alofa
Trinidad and Tobago	Port of Spain
Tunisia	Tunis
Turkey	Ankara
Uganda	Kampala
U.S.S.R.	Moscow
United Arab Emirates	Abu Dhabi
United Kingdom	London
U.S.A.	Washington D.C.
Upper Volta	Ouagadougou
Uruguay	Montevideo
Vanuatu	Vila
Venezuela	Caracas

Nation	*Capital*
Vietnam	Hanoi
Yemen (North)	Sana
Yemen (South)	Aden
Yugoslavia	Belgrade
Zaire	Kinshasa
Zambia	Lusaka
Zimbabwe	Harare

Appendix III

Per Capita Gross National Product

Nation	Mean Per Capita GNP (US $) 1980
North Africa:	
Algeria	1580
Egypt	460
Libya	8210
Morocco	740
Sudan	370
Tunisia	1120
West Africa:	
Benin	250
Gambia	260
Ghana	400
Guinea	270
Guinea-Bissau	170
Ivory Coast	1060
Liberia	490
Mali	140
Mauritania	320
Niger	270
Nigeria	670
Senegal	430
Sierra Leone	250
Togo	340
Upper Volta	180

Nation	Mean Per Capita GNP (US $) 1980
East Africa:	
Burundi	180
Djibouti	420
Ethiopia	130
Kenya	380
Madagascar	290
Malawi	200
Mauritius	1040
Mozambique	250
Reunion	4180
Rwanda	210
Seychelles	1400
Tanzania	270
Uganda	290
Zambia	510
Zimbabwe	470
Central Africa:	
Angola	440
Cameroon	560
Central African Republic	290
Chad	110
Congo	630
Gabon	3280
Zaire	260

Nation	Mean Per Capita GNP (US $) 1980
Southern Africa:	
Botswana	720
Lesotho	340
Namibia	1220
South Africa	1720
Swaziland	650
South-west Asia:	
Bahrain	5460
Cyprus	2940
Iraq	2410
Israel	4170
Jordan	1180
Kuwait	17270
Oman	2970
Qatar	16590
Saudi Arabia	7370
Syria	1070
Turkey	1330
United Arab Emirates	15590
Yemen (North)	420
Yemen (South)	500

Nation	Mean Per Capita GNP (US $) 1980
South Asia:	
Afghanistan	170
Bangladesh	100
Bhutan	80
India	190
Maldives	200
Nepal	130
Pakistan	270
Sri Lanka	230
South-east Asia:	
Brunei	10680
Burma	160
Indonesia	380
Malaysia	1320
Philippines	600
Singapore	3820
Thailand	590
East Asia:	
China	230
Hong Kong	4000
Japan	8800
Korea (North)	1130
Korea (South)	1500
Mongolia	780

Nation	Mean Per Capita GNP (US $) 1980
North America:	
Canada	9650
United States	10820
Central America:	
Belize	1030
Costa Rica	1810
El Salvador	670
Guatemala	1020
Honduras	530
Mexico	1590
Nicaragua	660
Panama	1350
Caribbean:	
Bahamas	2780
Barbados	2400
Cuba	1410
Dominica	410
Dominican Republic	990
Grenada	630
Guadeloupe	3260
Haiti	260
Jamaica	1240
Martinique	4680
Netherlands Antilles	3540
Puerto Rico	2970

Nation	Mean Per Capita GNP (US $) 1980
St. Lucia	780
St. Vincent/Grenadines	490
Trinidad and Tobago	3390

South America:

Argentina	2280
Bolivia	550
Brazil	1690
Chile	1690
Colombia	1010
Ecuador	1050
Guyana	570
Paraguay	1060
Peru	730
Surinam	2360
Uruguay	2090
Venezuela	3130

Northern Europe:

Denmark	11900
Finland	8260
Iceland	10490
Ireland	4230
Norway	10710
Sweden	11920
United Kingdom	6340

Nation	Mean Per Capita GNP (US $) 1980
Western Europe:	
Austria	8620
Belgium	10890
France	9940
Germany (West)	11730
Luxembourg	12820
Netherlands	10240
Switzerland	14240
Eastern Europe:	
Bulgaria	3690
Czechoslovakia	5290
Germany (East)	6430
Hungary	3850
Poland	3830
Romania	1900
Southern Europe:	
Albania	840
Greece	3890
Italy	5240
Malta	2640
Portugal	2160
Spain	4340
Yugoslavia	2430
U.S.S.R.	4110

Nation	Mean Per Capita GNP (US $) 1980
Oceania:	
Australia	9100
Fiji	1690
French Polynesia	6350
New Zealand	5940
Papua New Guinea	650

Appendix IV

The Geological Time Scale (Approximate)

Million years ago (not consistent scale)	Period	Era	Life forms
	Quaternary: Recent, Pleistocene	CENOZOIC	
1	Tertiary: Pliocene, Miocene, Oligocene, Eocene, Palaeocene	CENOZOIC	MAMMALS
63			
135	Cretaceous	MESOZOIC	
181	Jurassic	MESOZOIC	BIRDS
230	Triassic	MESOZOIC	REPTILES
280	Permian	PALAEOZOIC	AMPHIBIANS
345	Carboniferous	PALAEOZOIC	
405	Devonian	PALAEOZOIC	FISHES
425	Silurian	PALAEOZOIC	
500	Ordovician	PALAEOZOIC	
600	Cambrian	PALAEOZOIC	
>600	Pre-Cambrian		

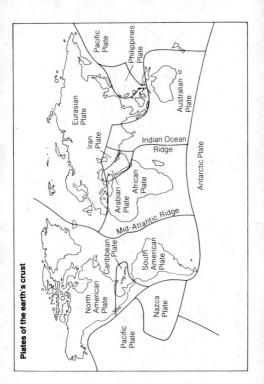

Plates of the earth's crust

Major ocean currents

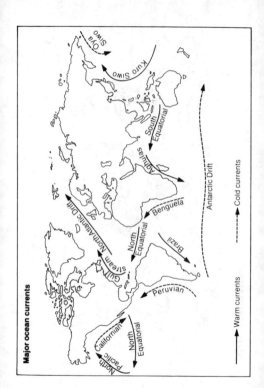

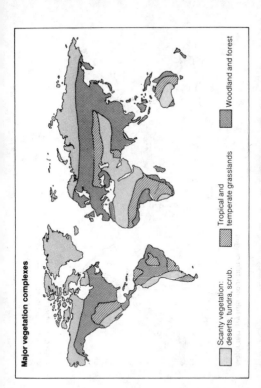

Major vegetation complexes

Scanty vegetation: deserts, tundra, scrub.

Tropical and temperate grasslands

Woodland and forest

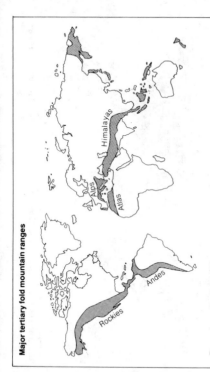

Major tertiary fold mountain ranges

Himalayas

Alps

Atlas

Rockies

Andes

The regions of tertiary folding are
also amongst the most active in terms
of earthquakes and volcanic events

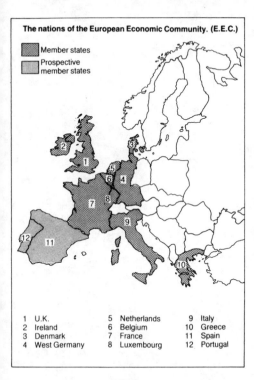

The nations of the European Economic Community. (E.E.C.)

Member states
Prospective member states

1 U.K.	5 Netherlands	9 Italy
2 Ireland	6 Belgium	10 Greece
3 Denmark	7 France	11 Spain
4 West Germany	8 Luxembourg	12 Portugal